Topics in Systems Engineering

Print ISSN: 2810-9090
Online ISSN: 2810-9104

Series Editors: Luigi Fortuna *(Università degli Studi di Catania, Italy)*
Arturo Buscarino *(Università degli Studi di Catania, Italy)*

The Series aims to cover a wide spectrum of Engineering topics but with a strong characteristic of interdisciplinarity using less technical items. It could be a series including thinking aspects, history topics linked with engineering topics from civil engineering to industrial engineering and so on. The contributions will cover a wide series of engineering topics and could be useful both for educational purposes and for research information. The idea is to combine technology, science, arts and social science with some emerging subjects.

Published:

Vol. 4:   *Nonlinear Dynamics of Laser*
by Riccardo Meucci and Jean-Marc Ginoux

Vol. 2:   *Modern Financial Engineering: Counterparty, Credit, Portfolio and Systemic Risks*
by Giuseppe Orlando, Michele Bufalo, Henry Penikas and Concetta Zurlo

Vol. 1:   *Underwater Labriform-Swimming Robot*
by Farah Abbas Naser, Mofeed Turky Rashid and Luigi Fortuna

Forthcoming:

Vol. 3:   *Fractional Discrete Chaos: Theories, Methods and Applications*
by Adel Ouannas, Iqbal M Batiha and Viet-Thanh Pham

Nonlinear Dynamics of Laser

Riccardo Meucci
Istituto Nazionale di Ottica, Italy

Jean-Marc Ginoux
Universite de Toulon, France

World Scientific

NEW JERSEY · LONDON · SINGAPORE · BEIJING · SHANGHAI · HONG KONG · TAIPEI · CHENNAI · TOKYO

Published by

World Scientific Publishing Co. Pte. Ltd.
5 Toh Tuck Link, Singapore 596224
USA office: 27 Warren Street, Suite 401-402, Hackensack, NJ 07601
UK office: 57 Shelton Street, Covent Garden, London WC2H 9HE

Library of Congress Control Number: 2023001512

British Library Cataloguing-in-Publication Data
A catalogue record for this book is available from the British Library.

Topics in Systems Engineering — Vol. 4
NONLINEAR DYNAMICS OF LASER

ISBN 978-981-127-251-6 (hardcover)
ISBN 978-981-127-252-3 (ebook for institutions)
ISBN 978-981-127-253-0 (ebook for individuals)

For any available supplementary material, please visit
https://www.worldscientific.com/worldscibooks/10.1142/13312#t=suppl

to the memory of Prof. Tito Arecchi...

"It was strange, in a way, because there were no ideas involved in the laser that weren't already known by somebody 25 years before lasers were discovered. The ideas were all there; just, nobody put it together."

— Charles H. Townes

In memoriam — Tito Arecchi (11 December 1933–15 February 2021)

Very recently a commemorative paper on Prof. Tito Arecchi appeared in *Chaos* [Meucci & Kurths (2022)]. The authors decided to reproduce it here entirely to honor his memory.

The nonlinear science community experienced a painful loss with the sudden death of our colleague and friend, Professor Tito Arecchi. Professor Arecchi was one of the 12 founding editors of the board of *Chaos* and later he became an Honorary Editor. He was very active and stimulating in forming and evolving this journal, resulting in such a serious and influential journal for Nonlinear Sciences and manifold applications. Tito Arecchi was a pioneer of nonlinear optics and laser physics as well as in nonlinear dynamics. His contributions have been so significant as to constitute milestones in the field of photon statistics and in that of nonlinear dynamics, not limited to lasers. In relation, we would like to emphasize the year 1965 as emblematic for his brilliant personality. He published two fundamental contributions: In the first one [Arecchi (1965)], he presented the first experimental evidence of the statistical difference between a laser and a random field obtained by photon statistics. In the second one, in collaboration with his first student Rodolfo Bonifacio, they derived the nonlinear equations which describe an electromagnetic pulse interacting self-consistently with an ensemble of two-level atoms under the assumption of a homogeneously broadened electric-dipole transition with two Bloch relaxation times T_2 ($\gamma_\perp = 1/T_2$) and T_1 ($\gamma_\| = 1/T_1$), and of a linear broadband loss mechanism [Arecchi & Bonifacio (1965)]. These equations are usually referred to as the Maxwell–Bloch equations but they should be referred to as the Arecchi–Bonifacio equations [McNeil (2015)]. In these equations, the Slowly Varying Envelope Approximation (SVEA) for the electromagnetic pulse was introduced for the first time. The Arecchi–Bonifacio equations are universally used to describe the dynamics of a single mode laser. Nowadays, it is well known that they are formally equivalent to those of the Lorenz model [Lorenz (1963)] and therefore, chaotic behavior is inherent in a laser [Haken (1975)]. However, we had to wait until 1982 to give an experimental confirmation using a single mode CO_2 laser with sinusoidal modulation of the cavity losses [Arecchi *et al.* (1982)]. This is due to the fact that in a large class of lasers, the so-called class B lasers, the macroscopic polarization evolves on fast time scales compared with the two other dynamical variables, i.e. the laser intensity, which is proportional to the photon number of the laser field mode, and the population inversion ($\gamma_\perp > k > \gamma_\parallel$), where k is the decay rate for the electric field. Few years later, Lorenz type chaos has been demonstrated for class C lasers, where the three decay rates are of the same order of magnitude [Weiss & Brock (1986)]. The above classification of lasers is another crucial contribution by Tito Arecchi [Arecchi *et al.* (1984)]. However, linking the scientific activities of Tito only to these aspects would be limiting. Tito has developed, with several colleagues, other important lines of research in the fields of complex systems, both from a theoretical and experimental point of view; among which we recall the optical vortices and their statistics, control and synchronization of chaos, multistability and even applications to neuroscience [Arecchi & Kurths (2009)]. It is important to emphasize that Riccardo Meucci *et al.* have recently revisited the rather simple laser model used in 1982 [Arecchi *et al.* (1982)]. They have highlighted new aspects on the relationship between multistability and dissipativity as well as its control [Meucci *et al.* (2022)]. Generalized multistability, another pioneering contribution by Tito, has become a focusing issue in many different fields as the numerous papers published in *Chaos* demonstrate [Feudel *et al.* (2018)]. In this latest period, when the pandemic has profoundly changed our lives, Tito's enthusiasm and passion for physics have not diminished, until few days before his departure, and no one who has interacted with him can forget it. This is the greatest legacy of him to science. He will be greatly missed by his many colleagues, former students and friends.

References

Meucci R. & Kurths J. (2022). In memoriam — Tito Arecchi (11 December 1933–15 February 2021), *Chaos*, 32, 080401

Arecchi F.T. (1965). Measurement of the statistical distribution of Gaussian and laser sources, *Physical Review Letters*, 15(24), pp. 912-916.

Arecchi F.T. & Bonifacio R. (1965). Theory of optical maser amplifiers, *IEEE Journal of Quantum Electronic*, 1(4), pp. 169-178.

McNeil B. (2015). Due credit for Maxwell-Bloch equations, *Nature Photonics*, 9, 207.

Lorenz E.N. (1963). Deterministic non-periodic flows, *J. Atmos. Sci.*, 20, pp. 130-141.

Haken H. (1975). Analogy between higher instabilities in fluids and lasers, *Physics Letters A*, 53(1), pp. 77-78.

Arecchi F.T., Meucci R., Puccioni G.P. & Tredicce J.R. (1982). Experimental evidence of subharmonic bifurcations, multistability, and turbulence in a Q-switched gas laser, *Physical Review Letters*, 49(17), pp. 1217-1220.

Weiss C.O. & Brock J. (1986). Evidence for Lorenz-type chaos in a laser, *Physical Review Letters*, 57(22), pp. 2804-2806.

Arecchi F.T., Lippi G.L., Puccioni G. & Tredicce J.R. (1984). Deterministic chaos in laser with injected signal, *Opt. Commun.*, 51(5), pp. 308-314.

Arecchi F.T. & Kurths J. (2009). Introduction to focus issue: Nonlinear dynamics in cognitive and neural systems, *Chaos*, 19, 015101.

Meucci R., Ginoux J.M., Mehrabbeik M., Jafari S. & Sprott J.C. (2022). Generalized multistability and its control in a laser, *Chaos*, 32(8), 083111.

Feudel U., Pisarchik A.N. & Showalter K. (2018). Multistability and tipping: From mathematics and physics to climate and brain — Minireview and preface to the focus issue, *Chaos*, 28, 033501 (8 pages).

Acknowledgments

The authors would like to thank Professors Luigi Fortuna, Guanrong (Ron) Chen, Juergen Kurths, Samuel Zambrano, Stefano Boccaletti, Julien C. Sprott, Angelo di Garbo, Jesus Seoane, Ines P. Marino, Francesco Marino, Stefano Euzzor, Eugenio Pugliese, Miguel Angel Sanjuan, Antonio Lapucci, Marco Ciofini, Massimiliano Locatelli, Mahtab Mehrabbeik, Sajad Jafari, Kais Al Naimee ... To our families and wives, who have supported us, we would like to extend our gratitude and love.

Contents

List of Figures

List of Tables

Chapter 1

Laser Physics and Laser Instabilities

The purpose of this chapter is to provide an introduction to the field of laser theory and laser instabilities. We briefly derive and discuss one of the most popular theoretical models: that for the homogeneously broadened single mode, laser. For a continuous wave (c.w.) pumped laser it seems natural to expect a steady output, and in fact, many lasers can be operated in a very stable way. However, for specific conditions and for certain types of lasers, the output may vary in time. In this case, we say that the laser has developed an instability. Instabilities induced by the nonlinearity of the interaction between radiation and matter have been carefully investigated both theoretically and experimentally.

1.1 Introduction

A laser consists of a dielectric material confined between, two reflecting mirrors acting as an optical cavity. The energy spectrum of the dielectric must contain two atomic or molecular energy levels whose populations are inverted with respect to the equilibrium condition. Different mechanisms may be responsible for achieving population inversion in the medium, such as an electrical discharge, optical pumping or chemical energy. When population inversion is established, atoms in higher level emit photons corresponding to energy difference between the two levels, by spontaneous and stimulated emission. The photons are reflected between the mirrors many times, thus stimulating the intense electromagnetic field characteristic for the laser. It is important to consider that before the laser process starts, the excited atoms in the laser resonator emit light spontaneously into all possible directions. On the other hand, an optical resonator such as a Fabry–Pérot interferometer, strongly discriminates among the possible frequencies giving preference to "modes" which have relatively long lifetimes. In other terms, only those light waves (modes) which can stay long enough in the resonator cause significant stimulated emission of the atoms. If we limit our considerations to the axial modes we observe that a resonator can support only modes for which $n\lambda/2 = L$ where λ is the wavelength in the medium, L is the distance between the mirrors and n is an integer number. The axial modes have frequency separation $\Delta f = c/2L$.

1.2   Semiclassical theory

In this section, the semiclassical laser theory is derived following the approaches of Lamb and Haken [Sargent *et al.* (1974); Haken (1985)]. In the semiclassical. framework, the light field is treated as a classical variable neglecting its operator character while atoms are treated using the formalism of quantum mechanics. We restrict our investigation to the case where the active material consists of two level atoms with a homogeneously broadened line. Usually in laser theory we consider the scalar transverse electric field strength $E(z,t)$ and we expand it into spatial modes $u_\lambda(z)$ determined by the resonator. For a Fabry–Pérot resonator the electric field can then be represented as a superposition of standing waves $u_\lambda(z)$:

$$E\left(z,t\right) = \sum_\lambda E_\lambda\left(z\right) u_\lambda\left(z\right) \tag{1.1}$$

where $u_\lambda\left(z\right) = \sin\kappa_\lambda$, which can be written as:

$$E\left(z,t\right) = \sqrt{\frac{\hbar\omega_0}{2\varepsilon_0 V}}\left[b(t)e^{-i(\omega_0 t - \kappa z)} + b^*(t)e^{i(\omega_0 t - \kappa z)}\right]$$

Thus we have:

$$\left|E\left(z,t\right)\right|^2 = \frac{\hbar\omega_0}{2\varepsilon_0 V}\left|b(t)\right|^2$$

The mode amplitudes $E_\lambda(t)$ are decomposed into their positive and negative frequency parts as follows:

$$E_\lambda\left(t\right) = E_\lambda^{(+)}e^{-i\omega_\lambda t} + E_\lambda^{(-)}e^{-i\omega_\lambda t} \tag{1.2}$$

where ω_λ is the frequency of the mode λ of the empty cavity. In our treatment we introduce the following normalization for $E^{(\pm)}$:

$$E_\lambda^{(-)} = \sqrt{\frac{\hbar\omega_\lambda}{2\varepsilon_0 V}}b^*(t)$$

$$E_\lambda^{(+)} = \sqrt{\frac{\hbar\omega_\lambda}{2\varepsilon_0 V}}b(t)$$

(1.3)

where V is the volume of the laser cavity. From a simple energy calculation we can show that b^*b is equal to the number of photons in the cavity. The energy density W of the laser radiation, in the case of single mode laser, is

$$W = \frac{1}{2}\varepsilon_0 |E(z,t)|^2 + \frac{1}{2}\mu_0 |H(z,t)|^2 = \varepsilon_0 |E(z,t)|^2$$
$$= 2\varepsilon_0 \left|E_\lambda^{(\pm)}(t)\right|^2 = \frac{\hbar\omega_\lambda}{V}b_\lambda^* b_\lambda$$

(1.4)

Integrating over the mode volume V and dividing by the energy of the single photon $\hbar\omega_\lambda$, we observe that using our normalization $b_\lambda^* b_\lambda$ represents the semiclassical photon number in the mode λ. The amplitudes $E_\lambda^{(+)}(t)(b(t))$ and $E_\lambda^{(-)}(t)(b^*(t))$ are time dependent complex functions but with a time dependence much slower than that of their accompanying exponential functions.

In the fully quantum-mechanical treatment we have the correspondence

$$b_\lambda^*(t) \rightarrow b_\lambda^+ \qquad b_\lambda(t) \rightarrow b_\lambda$$

(1.5)

where b^+ and b are the creation and annihilation operators, respectively, of photons in the field mode λ. The operators band $b+$ obey the commutation relation $[b, b^+] = 1$.

We assume that the field-matter coupling interaction can be described in the electric dipole moment approximation. The interaction Hamiltonian can be given by

$$H_{int} = -e\hat{x}.E(z,t),$$

(1.6)

where we consider the electric field $E(z,t)$ propagating along the laser axis z and polarized perpendicular to it in a direction we will call x. ex is electric dipole displaced by x. The total Hamiltonian is given by

$$H = H_0 - e\hat{x}.E(z,t),$$

(1.7)

where H_0 is the unperturbed Hamiltonian. The density operator, which characterizes the statistical behavior of atom-field system evolves in time following the equation

$$\dot{\rho}^{(\mu)} = -\frac{i}{\hbar} \left[H, \rho^{(\mu)} \right] \tag{1.8}$$

As we consider only a two-level system, the density operator can be expanded in terms of the eigenstates $|1\rangle$ and $|2\rangle$ of the Hamiltonian operator H_0.

$$H_0|1\rangle = E_1|1\rangle$$

$$H_0|2\rangle = E_2|1\rangle \qquad E_2 - E_1 = \hbar\omega_0 \tag{1.9}$$

$$\langle i|j\rangle = \delta_{ij}$$

We assume that there is no permanent dipole moment in the ground and excited states $\langle 1|\hat{x}|1\rangle = \langle 2|\hat{x}|2\rangle = 0$. The phases of the eigenstates $|1\rangle$ and $|2\rangle$ are also chosen so that $\langle 1|e\hat{x}|2\rangle \equiv ex_{12} \equiv \langle 2|e\hat{x}|1\rangle$ is a real number. With these assumptions the evolution equation for the elements ρ_{ij} are:

$$\dot{\rho}_{21}^{(\mu)} = -i\omega_0\rho_{21}^{(\mu)} + i\left(\frac{e}{\hbar}\right) x_{12} E\left(z,t\right) \left(\rho_{22}^{(\mu)} - \rho_{11}^{(\mu)}\right) - \gamma_\perp \rho_{21}^{(\mu)} \tag{1.10}$$

$$\dot{\rho}_{22}^{(\mu)} - \dot{\rho}_{11}^{(\mu)} = i\frac{2}{\hbar}ex_{12}E\left(z,t\right)\left(\rho_{12}^{(\mu)} - \rho_{21}^{(\mu)}\right) + \gamma_\parallel \left[d_0 - \left(\rho_{22}^{(\mu)} - \rho_{11}^{(\mu)}\right)\right] \tag{1.11}$$

The last two tenus in the above equations are added in a phenomenological way in order to describe damping processes of the off-diagonal elements $(-\gamma_\perp \rho_{21}^{(\mu)})$ and of the diagonal elements $(-\gamma_\parallel(\rho_{22}^{(\mu)} - \rho_{11}^{(\mu)}))$ due to such processes being spontaneous emission or collisions and in order to describe pumping mechanism $(\gamma_\parallel d_0)$ that creates the inversion. The damping rate $\gamma_\perp$ is often referred to as the dipole's dephasing rate in order to distinguish it from an energy decay rate. In a gas medium, elastic collisions have the effect of interrupting the phase of the electron's oscillations. To account for inelastic collisions and spontaneous emission whose effect is to depopulate the energy levels 2 and 1 into other unspecified energy levels of the atom, we introduce the relaxation rate $\gamma_\parallel$. As an exercise one can show that $\gamma_\perp$ is related to the width (Half Width at Half Maximum: H.W.H.M.) of the Lorentzian lineshape function which describes the gain of a homogeneously broadened atomic or molecular transition. A Lorentzian lineshape function $L(\nu) = \dfrac{\delta\nu_0/\pi}{(\nu - \nu_0)^2 + \delta'\nu_0^2}$ is completely defined by assigning $\delta\nu_0 = \dfrac{\gamma_\perp}{2\pi}$ which is the half width at half maximum and the resonant frequency ν_0. Damping constants $\gamma_{\parallel}$ and $\gamma_{\perp}$ have the some meaning as the inverse of the relaxation times T_1 and T_2 of the Bloch equations of the spin resonance theory. Using the field expansion (1.1) and the rotating wave approximation (R.W.A.), Eq. (1.10) becomes

$$\dot{\rho}_{21}^{(\mu)} = -i\omega_0 \rho_{21}^{(\mu)} - \frac{iex_{21}}{\hbar} d^{(\mu)} \sum_{\lambda} E_{\lambda}^{+}(t) e^{-i\omega_\lambda t} u_\lambda(z_\mu) - \gamma_{\perp}\rho_{21}^{(\mu)} \qquad (1.12)$$

Application of R.W.A. means that we retain only those terms evolving in time at about $e^{-i\omega_\lambda t}$, the temporal evolution due to the first term $\rho_{21}(t) \propto e^{-i\omega_0 t}$. This approximation is justified since the laser process is important only for mode frequencies ω_λ close to the atomic transition frequency ω_0. It is useful to define new "slowly" varying variables $\tilde{\rho}_{21}$ and $\tilde{\rho}_{12}$ through the relations:

$$\rho_{21}(t) = \tilde{\rho}_{21}(t)e^{-i\omega t} \quad , \quad \rho_{12}(t) = \tilde{\rho}_{12}(t)e^{+i\omega t} \qquad (1.13)$$

For the case of a single mode laser, using the normalization (1.3) and the relation (1.13) we obtain:

$$\dot{\tilde{\rho}}_{21}^{(\mu)} = (-i(\omega_0 - \omega) - \gamma_{\perp})\, \tilde{\rho}_{21}^{(\mu)} - ig_{\mu\lambda}d^{(\mu)}b \qquad (1.14)$$

where

$$g_{\mu\lambda} = \frac{e}{\hbar}x_{12}u(z_\mu)\sqrt{\frac{\hbar\omega}{2\varepsilon_0 V}} = gu(z_\mu)$$

is the field-matter coupling constant.

For simplicity the coupling constant will be assumed to be independent on the position of the atom z_μ, so that $g_{\mu\lambda}, = g$. This approximation is nearly exact for a single mode ring cavity and gives little error for Fabry–Pérot laser near threshold. Making use of the same hypothesis we obtain the following equation for the population inversion $\rho_{22}^{(\mu)} - \rho_{11}^{(\mu)} = d^{(\mu)}$.

$$\dot{d}^{(\mu)} = \gamma_{\parallel}\left(d_0 - d^{(\mu)}\right) + 2ig\left(b\tilde{\rho}_{12}^{(\mu)} - b^*\tilde{\rho}_{21}^{(\mu)}\right) \qquad (1.15)$$

From the Maxwell equations, we obtain the one-dimensional wave equation for the electric field $E(z,t)$ driven by the polarization $P(z,t)$:

$$\frac{\partial^2 E}{\partial t^2} + 2k\frac{\partial E}{\partial t} - c^2\frac{\partial^2 E}{\partial z^2} = -\frac{1}{\varepsilon_0}\frac{\partial^2 P}{\partial t^2} \tag{1.16}$$

where k describes the losses of the electric field. Let us assume that the only cause of loss is associated with transmission through the mirrors of the optical cavity. In the bare cavity, the cavity intensity decays exponentially:

$$I_\lambda(t) = I_\lambda(0)\, e^{-\frac{c}{2L}\ln\frac{1}{R_1 R_2}t} \tag{1.17}$$

where L is the cavity length, R_1 and R_2 are the reflectivities of the cavity mirrors ($R_1, R_2 < 1$). Since the intensity is proportional to the square of the electric field amplitude $E_\lambda(t)$, it follows that the frequency spectrum associated with the time-dependent free field in the cavity is a Lorentzian with a bandwidth $\delta\nu_c = \frac{1}{2\pi}\frac{1}{2}\frac{c}{2L}\ln\frac{1}{R_1 R_2} = \frac{1}{2\pi}k$.

Frequently the resonances of a cavity are characterized by the dimensionless quality factor

$$Q = \frac{1}{2}\frac{\nu}{\delta\nu_c} \tag{1.18}$$

A high-Q cavity is one with low loss, while a low-Q cavity has a high loss rate. These results can be generalized to include other losses effect.

The macroscopic polarization $P(z,t)$ is related to the microscopic variables $\rho_{12}^{(\mu)}$ and $\rho_{21}^{(\mu)}$ by the following relation

$$P(z,t) = \frac{1}{V}\sum_\mu ex_{12}\left(\rho_{12}^{(\mu)} + \rho_{21}^{(\mu)}\right)\delta(z - z_\mu) \tag{1.19}$$

where δ is the Dirac's δ-function.

Introducing the field expansion (1.1) in the wave equation (1.16) we obtain for the amplitude $E(t)$ (as we consider only one mode the subscript λ will be neglected) the following equation:

$$\omega_\lambda^2 E(t) + \ddot{E}(t) + 2k\dot{E}(t) = -\frac{1}{\varepsilon_0}\ddot{P}(t) \tag{1.20}$$

where ω_λ is the frequency of the bare-cavity mode, and

$$P(t) = \frac{1}{V}\sum_\mu ex_{12}\left(\rho_{12}^{(\mu)} + \rho_{21}^{(\mu)}\right)u(z_\mu).$$

Applying the RWA and the Slowly Varying Amplitude Approximation (SVEA) to Eq. (1.20), we obtain

$$\dot{E}^{(+)}(t) = \left(-i(\omega_\lambda - \omega) - k\right) E^{(+)}(t) + \frac{i\omega_0}{2\varepsilon_0} P^{(+)}(t) \qquad (1.21)$$

where

$$P^{(+)}(t) = \frac{1}{V} \sum_\mu ex_{12} \tilde{\rho}_{21}^{(\mu)} u(z_\mu).$$

The SVEA approximation means that the field amplitude $E^{(+)}(t)$ evolves slowly on the time scale of a period of its accompanying exponential function. Therefore, we may assume that the temporal derivative of $E^{(+)}(t)$ is much smaller than $\omega E^{(+)}(t)$, i.e. $|\dot{E}^{(+)}(t)| \ll \omega |E^{(+)}(t)|$.

Using the normalization for electric field we finally obtain:

$$\dot{b}(t) = \left(-i(\omega_\lambda - \omega) - k\right) b(t) + ig \sum_\mu \tilde{\rho}_{21}^{(\mu)}. \qquad (1.22)$$

Introducing the collective variables $S(t) = \sum_\mu \tilde{\rho}_{21}^{(\mu)}(t)$ and $D(t) = \sum_\mu d^{(\mu)}$, assuming the resonance condition $\omega = \omega_\lambda = \omega_0$ one obtains the following equations (the so-called Maxwell–Bloch equations):

$$\dot{b} = -kb + igS \qquad (1.23)$$

$$\dot{S} = -\gamma_\perp S - igbD \qquad (1.24)$$

$$\dot{D} = \gamma_\parallel (D_0 - D) + 2ig (bS^* - b^*S) \qquad (1.25)$$

We briefly discuss the physical content of these equations. The right-hand side of Eq. (1.23) describes the causes of the temporal change of the field amplitude. The first term describes the damping of the field amplitude in the empty resonator, while the second describes how the dipole moment acts as a source for the field. In analogy to the field equation, the first term of Eq. (1.24) describes the damping of the dipole moment. The last term describes the interaction of the field mode with the atoms, which create a dipole moment. As we are dealing with two-level atoms, the energy flux between atoms and the field depends on the state of the atoms. If the populations are inverted, energy will be transferred from the atoms to the dipole moments. On the other hand, if the atoms are in their lower state, energy will be transferred from the field to the atoms by absorption.

In the equation for population inversion (Eq. (1.25)) the first term describes the relaxation of the inversion. $\gamma_{\parallel}$ is the relaxation rate, while D_0 represents the equilibrium inversion in the absence of field. The last term is proportional to the energy per second put into the atoms or extracted from the atoms caused by the coherent interaction between the dipole moment of the atoms and the field. The conceptual foundations of the semiclassical theory can be summarized in the scheme reported just below. An incident electromagnetic field interacts with a collection of microscopic dipoles and creates a macroscopic polarization. This polarization acts as a driving force for the field which interacts again with the microscopic dipoles. An external pump mechanism provides energy to the active medium in order to establish population inversion.

$$E\left(z,t\right) \xrightarrow[mechanics]{Quantum} \langle ex \rangle_{\mu} \xrightarrow[summation]{statistical} P\left(z,t\right) \xrightarrow[equations]{Maxwell's} E\left(z,t\right)$$

The semiclassical approach to the laser theory

1.3   Fundamental results of the semiclassical theory

The Maxwell–Bloch equations have a non-zero solution, if the pump parameter D_0 is larger than the threshold value

$$D_0 > D_{ths} = \frac{k\gamma_{\perp}}{g^2}. \tag{1.26}$$

The steady state photon number is given by

$$b^*b = \frac{\gamma_{\parallel}}{4k}\left(D_0 - D_{ths}\right). \tag{1.27}$$

The steady state population inversion which is clamped to the threshold value is given by $D_{ss} = \dfrac{k\gamma_{\perp}}{g^2}$. As a consequence we can find how b^*b depends on D_0 getting

$$b^*b = n_s\left(\frac{D_0}{D_{ths}} - 1\right). \tag{1.28}$$

Solving Eq. (1.28) for D_0 we obtain the well known expression for the saturation of the population inversion:

$$D_{ths} = \frac{D_0}{1 + \dfrac{4g^2 b^* b}{\gamma_\perp \gamma_\parallel}} = \frac{D_0}{1 + \dfrac{b^* b}{n_s}}. \tag{1.29}$$

where n_s is saturation photon number defined as

$$n_s = \frac{\gamma_\perp \gamma_\parallel}{4g^2} = \frac{\gamma_\parallel}{2G} \text{ where } G = \frac{2g^2}{\gamma_\perp}. \tag{1.30}$$

It should be noted that, if $\omega = \omega_\lambda = \omega_0$, S and b can be considered as real variables. In this case the related dynamical motion is confined to a three-dimensional phase space. However, by considering the different time scales of relaxation processes in this phase space, the dynamics may be limited further to a subspace, giving rise to a well established classification of lasers such as Class A, Band C lasers [Arecchi (1987)].

In the most familiar gas lasers (He-Ne, Ar^+, Kr) and in dye lasers the population and polarization variables decay much faster than that of the electric field, so the corresponding equations can be solved at equilibrium with the field and their quasi steady state solution can be substituted in the field equation. More precisely when $\gamma_\perp, \gamma_\parallel > k$ we can perform the adiabatic elimination of the two atomic variables. The field equation becomes:

$$\dot{b} + kb = ig\left(-\frac{igb}{\gamma_\perp} D_{ths}\right). \tag{1.31}$$

using the relation (1.29) we have

$$\dot{b} + kb = \frac{g^2}{\gamma_\perp} b \left(\frac{D_0}{1 + \dfrac{4g^2 b^* b}{\gamma_\perp \gamma_\parallel}}\right). \tag{1.32}$$

Near threshold where $b^* b \ll n_s$ Eq. (1.32) can be written approximately as

$$\dot{b} = \left(-k + \frac{g^2}{\gamma_\perp} D_0\right) b - \frac{4g^4}{\gamma_\perp^2 \gamma_\parallel} D_0 |b|^2 b. \tag{1.33}$$

The first term in the brackets on the rhs stems from the cavity losses, the second positive term describes the gain of the unsaturated inversion. The last term describes the saturation effect of the laser process. In some other lasers, as e.g. ruby, Nd, CO_2 and semiconductor lasers, only the polarization decay rate is sufficiently larger compared to the other two rates. In this case, only the polarization can be adiabatically eliminated and the resulting laser dynamics is described by the "rate equations":

$$\dot{b} + kb = \frac{g^2}{\gamma_\perp} bD \tag{1.34}$$

$$\dot{D} + \gamma_\| (D - D_0) = -\frac{4g^2}{\gamma_\perp} |b|^2 D \tag{1.35}$$

Considering that in our notation $b^*b = n =$ photon number, Eqs. (1.34)–(1.35) can be rewritten as:

$$\dot{n} + 2kn = GnD \tag{1.36}$$

$$\dot{D} + \gamma_\| (D - D_0) = -2GnD \tag{1.37}$$

Let us now consider the stability of the c.w. (continuous wave) solution of the rate equations (1.36)–(1.37). It is convenient to introduce the normalized quantities

$$\hat{n} = \frac{n}{n_{cw}} , \quad \hat{D} = \frac{D}{D_{ths}} \tag{1.38}$$

and the normalized pump parameter

$$\Lambda = \frac{D_0 - D_{ths}}{D_{ths}} \tag{1.39}$$

Using these normalized quantities Eqs. (1.36)–(1.37) become

$$\dot{\hat{n}} + 2k\hat{n} = 2k\hat{n}\hat{D} \tag{1.40}$$

$$\dot{\hat{D}} + \gamma_\| \hat{D} = \gamma_\| (\Lambda + 1) - \gamma_\| \Lambda \hat{n}\hat{D} \tag{1.41}$$

In order to check the stability of the stationary solution $\hat{D} = \hat{n} = 1$ we introduce small deviations $\delta\hat{n}$ and $\delta\hat{D}$ from the steady state solution. The linearized equations for $\delta\hat{n}$ and $\delta\hat{D}$ can be written as:

$$\begin{pmatrix} \dot{\delta\hat{n}} \\ \dot{\delta\hat{D}} \end{pmatrix} = \begin{pmatrix} 0 & +2k \\ -\gamma_\|\Lambda & -\gamma_\| (1 + \Lambda) \end{pmatrix} \begin{pmatrix} \delta\hat{n} \\ \delta\hat{D} \end{pmatrix} \tag{1.42}$$

These equations have a solution given by

$$\begin{pmatrix} \delta \hat{n}(t) \\ \delta \hat{D}(t) \end{pmatrix} = \begin{pmatrix} \delta \hat{n}(0) \\ \delta \hat{D}(0) \end{pmatrix} e^{\lambda t}$$

where the eigenvalues of the stability matrix are the roots of the polynomial

$$P(\lambda) = \lambda^2 + \gamma_{\parallel}(1 + \Lambda)\lambda + 2k\gamma_{\parallel}\Lambda \tag{1.43}$$

These eigenvalues are given by

$$\lambda_{12} = -\frac{\gamma_{\parallel}}{2}(1 + \Lambda) \pm i\sqrt{2k\gamma_{\parallel}\Lambda - \left[\frac{\gamma_{\parallel}}{2}(1 + \Lambda)\right]^2} \tag{1.44}$$

One observes that the steady state solution is always stable. The intensity oscillates about the steady state value and approaches $n_c w$ at the exponential rate $\frac{\gamma_{\parallel}}{2}(1 + \Lambda)$. The relaxation oscillations occur at a frequency

$$\omega_r = \sqrt{2k\gamma_{\parallel}\Lambda - \left[\frac{\gamma_{\parallel}}{2}(1 + \Lambda)\right]^2} \tag{1.45}$$

When the three decay rates for the polarization, population and field are of the same order of magnitude (class C lasers) all three equations are essential.

We investigate now the stability of the c.w. solution of the single mode laser equations (1.23)–(1.25). It is useful to introduce the following normalized variables:

$$\hat{b} = \frac{b}{b_{cw}}, \ \hat{S} = \frac{S}{S_{cw}}, \ \hat{D} = \frac{D}{D_{cw}} \tag{1.46}$$

A little algebra transforms the laser equations into the following normalized form:

$$\dot{\hat{b}} + k\hat{b} = k\hat{S} \tag{1.47}$$

$$\dot{\hat{S}} + \gamma_{\perp}\hat{S} = \gamma_{\perp}\hat{b}\hat{D} \tag{1.48}$$

$$\dot{\hat{D}} + \gamma_{\parallel}\hat{D} = \gamma_{\parallel}(\Lambda + 1) - \frac{1}{2}\gamma_{\parallel}\Lambda\left(\hat{S}^*\hat{b} + \hat{S}\hat{b}^*\right) \tag{1.49}$$

The c.w. solution reads, of course,

$$\hat{b} = \hat{S} = \hat{D} = 1 \tag{1.50}$$

Introducing the small deviations $\delta\hat{b}, \delta\hat{S}$ and $\delta\hat{D}$ from the c.w. solution, the linearized equations for $\delta\hat{b}, \delta\hat{S}$ and $\delta\hat{D}$ are:

$$\begin{pmatrix} \delta\dot{\hat{b}} \\ \delta\dot{\hat{S}} \\ \delta\dot{\hat{D}} \end{pmatrix} = \begin{pmatrix} -k & +k & 0 \\ \gamma_\perp & -\gamma_\perp & +\gamma_\perp \\ -\gamma_\parallel\Lambda & -\gamma_\parallel\Lambda & -\gamma_\parallel\Lambda \end{pmatrix} \begin{pmatrix} \delta\hat{b} \\ \delta\hat{S} \\ \delta\hat{D} \end{pmatrix} \tag{1.51}$$

Solutions of the form

$$\begin{pmatrix} \delta\hat{b} \\ \delta\hat{S} \\ \delta\hat{D} \end{pmatrix} = \begin{pmatrix} \delta\hat{b}_0 \\ \delta\hat{S}_0 \\ \delta\hat{D}_0 \end{pmatrix} e^{\lambda t}$$

lead to the cubic equation

$$\lambda^3 + \lambda^2\left(k + \gamma_\perp + \gamma_\parallel\right) + \lambda\left(k\gamma_\parallel + \gamma_\perp\gamma_\parallel + \gamma_\perp\gamma_\parallel\Lambda\right) + 2\gamma_\parallel\gamma_\perp k\Lambda = 0 \tag{1.52}$$

It follows from the Hurwitz criterium that the solutions of Eq. (1.52) are stable, i.e. all eigenvalues have negative real parts if either $\gamma_\perp + \gamma_\parallel \geq k$ or $\gamma_\perp + \gamma_\parallel < k$ and $\Lambda \leq \Lambda_c$. Here, Λ_c is given by

$$\Lambda_c = \frac{\left(\gamma_\parallel + \gamma_\perp + k\right)\left(\gamma_\perp + k\right)}{\left(k - \gamma_\parallel - \gamma_\perp\right)\gamma_\perp} \tag{1.53}$$

The solutions are unstable if

$$\gamma_\parallel + \gamma_\perp < k \text{ and } \Lambda > \Lambda_c \tag{1.54}$$

Thus in order to get unstable solutions for the single mode laser the cavity losses must exceed the sum of the losses of the polarization and the inversion (bad cavity condition). Furthermore, the pump strength must be very high in order to fulfill the condition $\Lambda > \Lambda_c$.

It was demonstrated by H. Haken [Haken (1975)] that Eqs. (1.47)–(1.49) are equivalent to the Lorenz equations which provide a simple model for convective turbulence in fluid-dynamics [Schuster (1988)]. Lorenz considered the problem of convection instability or Benard instability i.e. the problem of a fluid layer heated from below. The motion of the fluid is described by the Navier–Stokes equations which are nonlinear, partial differential equations. The Lorenz model comes from a truncation of a Fourier expansion of the velocity and temperature fields of the fluid. The Lorenz equations have the following form:

$$\dot{x} = \sigma y - \sigma x \tag{1.55}$$

$$\dot{y} = -xz + rx - y \tag{1.56}$$

$$\dot{z} = xy - bz \tag{1.57}$$

where σ is the Prandtl number, $r = \dfrac{R}{R_c}$ where R is Rayleigh number and R_c is the critical Rayleigh number for the onset of convection. A numerical analysis of this set of nonlinear differential equations shows that its variables can exhibit an irregular motion which is called "chaotic" when r is above a threshold value r_c.

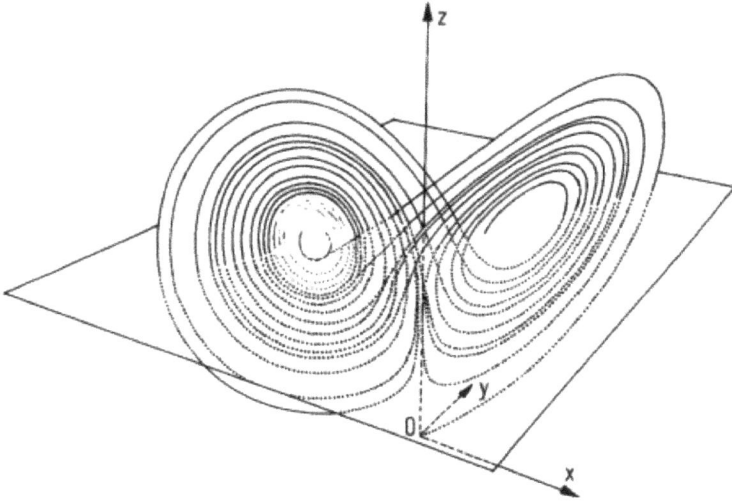

Figure 1.1. The Lorenz attractor, after a computer calculation by Lanford (1977). See [Schuster (1988)].

Figure 1.1 shows the trajectory generated by the Lorenz equations for $r = 28$, $\sigma = 20$ and $b = \dfrac{8}{3}$. One finds what the trajectory is attracted to a bounded region in phase space and it makes one loop to the right, then few loops to the left, then to the right and so on ... Another peculiarity of the "chaotic" solution is that if another adjacent initial condition is taken, the new solution exponentially deviates from the old (sensitive dependence on the initial condition).

As introduced in this paragraph the Lorenz equations play a crucial role in laser physics because they are equivalent to single mode laser equations. Introducing the transformations

$$\tau = \gamma_\perp t \ , \ x = \sqrt{\Lambda \frac{\gamma_\parallel}{\gamma_\perp}} \hat{b} \ , \ y = \sqrt{\Lambda \frac{\gamma_\parallel}{\gamma_\perp}} \hat{S} \text{ and } z = \Lambda + 1 - \hat{D} \qquad (1.58)$$

the laser Eqs. (1.47)–(1.49) are transformed to the Lorenz Eqs. (1.55)–(1.57) if σ, b, and r are defined as follows:

$$\sigma = \frac{k}{\gamma_\perp} \ , \ b = \frac{\gamma_\parallel}{\gamma_\perp} \ , \ r = \Lambda + 1 \ , \ \Lambda = \frac{D_0}{D_{ths}} - 1 \qquad (1.59)$$

Experimental evidence of chaotic instabilities showing characteristics similar to Lorenz chaos was found in the far infrared NH3 ammonia laser [Weiss & Brock (1986)] which is a class C laser. Chaos can also be observed in lasers by other mechanisms. One idea was to use class B lasers, where the atomic polarization can be adiabatically eliminated, but at the same time to introduce some modulation on a given parameter or to introduce optical or electroptical feedback schemes so that the number of variables remains (at least) three. In the literature [Abraham *et al.* (1988)] different configurations have been treated, such as

(a) time dependent modulation of the cavity losses,

(b) time dependent modulation of the population inversion,

(c) injection of a coherent electric field into the cavity of a laser,

(d) electroptic feedback,

(e) laser with saturable absorber.

A loss modulated single mode CO_2 laser was one of the first laser devices for which chaotic dynamics was investigated in a detailed way [Arecchi *et al.* (1982)]. The experiments were performed using a single mode CO_2 laser with an intracavity electroptic modulator allowing modulation of the cavity losses. The intensity decay rate can be expressed as follows:

$$K\left(t\right) = \frac{c}{2L}\left[2T + (1 - 2T)\sin^2 \pi \frac{V(t)}{V_\lambda}\right] \tag{1.60}$$

where L is the cavity length, T is the total transmission coefficient per pass of the cavity, V_λ is the $\lambda/2$ voltage which is a characteristic parameter of the EOM used and $V(t) = V_0 + V_1 \sin 2\pi ft$ the voltage applied to the modulator (V_0 is a suitable bias voltage). The modulation frequency f must be selected in a range close to the value of the relaxation oscillations estimated from the linear stability analysis. Considering that $V(t) < V_\lambda$ the above expression for the cavity losses can be approximated as $K(t) \simeq K_0(1 + m \sin 2\pi ft)$.

Acknowledgments

The author is indebted for fruitful conversations with Prof. N.B. Abraham. A debt of gratitude is devoted to Prof. F.T. Arecchi for introducing the author to the field.

1.4   References

Sargent M., Scully M.O. & Lamb W.E. (1974). *Laser Physics* (Addison-Wesley).

Haken H. (1985). *Light Volume 2: Laser Light Dynamics* (North Holland).

Arecchi F.T. (1987). *Instabilities and Chaos in Quantum Optics*, eds. F.T. Arecchi and R.O. Harrison (Springer, Berlin).

Haken H. (1975). Analogy between higher instabilities in fluids and lasers, *Physics Letters A*, 53(1), pp. 77-78.

Weiss C.O. & Brock J. (1986). Evidence for Lorenz-type chaos in a laser, *Physical Review Letters*, 57(22), pp. 2804-2806. (See also: Weiss C.O., Abraham N.B., Hübner U. (1988). *Physical Review Letters*, 61(14), pp. 1587-1590).

Abraham N.B., Narducci L.M. & Mandel P. (1988). *Progress in Optics, Volume XXV*, ed. H. Haken (Elsevier). (See also articles in "Selected papers on Optical Chaos," SPIE Milestone Series Vol. MS75, eds. F.T. Arecchi and R.O. Harrison, SPIE Optical Engineering Press (1993)).

Schuster H.O. (1988). *Deterministic Chaos* (VCH Weinheim).

Arecchi F.T., Meucci R., Puccioni G.P. & Tredicce J.R. (1982). Experimental evidence of subharmonic bifurcations, multistability, and turbulence in a Q-switched gas laser, *Physical Review Letters*, 49(17), pp. 1217-1220.

Chapter 2

Generalized Multistability and its Control in a Laser

In this chapter, we revisit the model of the laser with cavity loss modulation from which evidence of chaos and generalized multistability was discovered in 1982. Multistability refers to the coexistence of two or more attractors in nonlinear dynamical systems. Despite its relative simplicity, the adopted model shows us how the multistability depends on the dissipation of the system. The model is then tested under the action of a secondary sinusoidal perturbation, which can remove bistability when a suitable relative phase is chosen. The surviving attractor is the one with less dissipation. This control strategy is particularly useful when one of the competing attractors is a chaotic attractor.

We revisit the dynamics of a simple model used to describe the first experimental evidence of chaos in a modulated laser. This pioneering experiment had an enormous impact on the scientific community, considering that a laser could also emit in a chaotic way while retaining its optical coherence properties. Lasers, particularly class B-lasers, like the CO_2 and later semiconductor lasers, became reliable devices for studying chaos and generalized multistability. Nowadays, the latter phenomenon is widely investigated in the most diverse fields, sharing the possibility of jumping between the different attractors using small perturbations. The model has been explored in terms of its dissipativity. A novel aspect of the present investigation is the stability analysis in an increased dimension phase space allowing analytic treatment.

2.1   Introduction

Exactly forty years have passed since the pioneering experiment on deterministic chaos and generalized multistability in a CO_2 laser with periodic modulation of the cavity losses [Arecchi *et al.* (1982)]. These two issues have profoundly influenced and motivated research in fields different from that of laser physics. Let us consider multistability, that is, the coexistence of different stable states in nonlinear systems (for two review papers on the subject, see [Feudel (2008); Feudel *et al.* (2018)]). This means that a dissipative dynamical system can have more solutions for equal values of the control parameters depending only on the values of initial conditions. The set of initial conditions (more precisely, the closure of it) leading, in the long term limit, to a given attractor is called the basin of attraction whose structure can be fractal. The complicated structure of basin boundaries in multistable systems determines their sensitivity to noise and periodic perturbations. This makes them attractive for controlling techniques allowing the switching from one attractor to another one. Many dynamical systems exhibit multistability, including laser physics [Saucedio *et al.* (2003); Pisarchik *et al.* (2011)], neuron models [Schwartz *et al.* (2012)], chemical reactions [Ryashko (2018)], climate systems [Lucarini & Bodai (2017)], biological and ecological ones [May (1977)].

In the 1982 seminal paper [Arecchi *et al.* (1982)], a simple two-level laser model was used. A few years later, a five-dimensional model was introduced accounting for the interaction between the electromagnetic field and a molecular model where the two lasing levels are coupled to two rotational manifolds. Using the center manifold theory, it is possible to reduce the five-dimensional model to a two-dimensional model adding suitable nonlinear corrections as demonstrated by Ciofini et al [Ciofini *et al.* (1993)]. However, considering that the key nonlinearity is the same in the two models, it often is preferable to use the two-level model. Very recently, a simple three-dimensional laser model was proposed to investigate the instabilities of the laser with feedback. Such a model possesses the minimal and essential nonlinearities as the Roessler, Lorenz, Chua, and Chen models [Meucci *et al.* (2021); Ricci *et al.* (2021)]. The two-level model that we use here simply derives from it by eliminating the feedback equation and introducing a sinusoidal modulation of the cavity losses parameter, and recapturing the basic scheme of the one introduced in [Arecchi *et al.* (1982)]. In such a case, a certain flexibility is used in the parameter gamma (γ), which accounts for the relaxation rate of population inversion.

An increasing interest has been posed on generalized multistability and its control, considering the possibility of using small perturbations to select one of the competing solutions as demonstrated by Goswami and Pisarchick [Goswami & Pisarchick (2008)]. Here, we consider the key role of the phase difference between the main driving frequency responsible for chaos and multistability(f_{mod}) and a secondary sinusoidal perturbation for its control (f_{pert}). The attention is here focused on the resonant case where $f_{mod} = f_{pert}$ [Meucci *et al.* (1994, 2016)]. This chapter is organized as follows: In Section 2.2, we introduce the two-level laser model and its time rescaled version. Numerical results showing evidence of generalized bistability are here presented. In Section 2.3, we transform this two-level non-autonomous model into an autonomous four-dimensional dynamical system that enables us to provide a mathematical analysis and confirm the numerical results obtained. In Section 2.4, control of bistability is obtained by introducing a secondary sinusoidal perturbation adjusting the relative phase.

2.2   Two-level non-autonomous laser model

Starting from the seminal works of Arecchi *et al.* [Arecchi *et al.* (1982, 1986); Arecchi (1987)], we propose to analyze the following *two-level laser model*:

$$\dot{x} = -k_0 x \left[1 + k_1 \left(B_0 + m \sin(2\pi f_{mod} t) \right)^2 - y \right],$$

$$\dot{y} = -\gamma y - \frac{2k_0}{\alpha} xy + \gamma p_0. \tag{2.1}$$

where the fast variable x is the laser intensity with a time dependent decay rate $k(t)$ given by $k_0 \left[1 + k_1 \left(B_0 + m \sin(2\pi f_{mod} t) \right)^2 \right]$, where k_0 is the non-modulated cavity loss parameter and k_1 accounts for its modulation depth. $B_0 + m \sin(2\pi f_{mod} t)$ is the applied modulation signal consisting of a bias value B_0 summed to a sinusoidal signal with amplitude m and modulation frequency f_{mod}. The slow variable y is the population inversion with a decay rate γ, while the parameter y_0 is the population inversion imposed by the pumping process. The adopted normalization is such that the original equations in Ref. [Arecchi *et al.* (1982)] can be re-obtained considering $\alpha = 2k_0/\gamma$ and a threshold inversion $y_{thres} = k_0/G$ where G is the field-matter coupling constant that appears in Ref. [Arecchi *et al.* (1982)].

2.2.1 Rescaled form

We propose the following change of variables and parameters to recast Eq. (2.1) in a rescaled form. Let us suppose: $t \rightarrow \frac{t}{k_0}$ and

$$k = k_1, \ f'_{mod} = \frac{f_{mod}}{k_0}, \ \gamma' = \frac{\gamma}{k_0}, \ \alpha' = \frac{2}{\alpha}.$$

By dropping the ' for these parameters, the two-level model (2.1) now reads:

$$\frac{dx}{dt} = -x \left[1 + k \left(B_0 + m \sin(2\pi f_{mod}t) \right)^2 - y \right],$$

$$\frac{dy}{dt} = -\gamma y - \alpha xy + \gamma p_0,$$

(2.2)

where $k = 12$, $B_0 = 0.1215$, $m = 0.02$, $f_{mod} = 0.005$, $\gamma = 0.0025$, $\alpha = 0.002$ and $p_0 = 1.252$. In the following, we will use B_0 as the bifurcation parameter. Numerical investigations on the non-autonomous dynamical system (2.2) enable highlighting generalized bistability and its control. After presenting these numerical results, the two-level laser model (2.2) will be transformed into an autonomous four-dimensional dynamical system that will allow its mathematical analysis.

2.2.2 Jacobian matrix

The phenomenon of generalized multistability, that is, the coexistence of attractors for the same parameter values, is related to the phenomenon of crises [Grebogi *et al.* (1983)]. The presence of different crises in dynamical systems depends on the amount of dissipation [Feudel *et al.* (1996)]. Although an attempt to relate bistability and dissipation has already been made in this system (see Ref. [Meucci *et al.* (1988)]), here we want to trace it starting from the first principles. Dissipativity is related to the Jacobian matrix of the model [Sprott (2003); Ott (1993)].

The Jacobian matrix of the rescaled two-level model (2.2) reads:

$$J = \begin{pmatrix} y - k \left(B_0 + m \sin(2\pi f_{mod}t) \right)^2 - 1 & x \\ -\alpha y & -\gamma - \alpha x \end{pmatrix}$$

(2.3)

So, the trace of the Jacobian which represents the dissipation rate of the rescaled two-level model (2.2) reads:

$$Tr(J) = y - k\left(B_0 + m\sin(2\pi f_{mod}t)\right)^2 - 1 - \gamma - \alpha x. \qquad (2.4)$$

In Figure 2.1a, the bifurcation diagram of system (2.2), i.e. the maxima of the solution $x(t)$ as a function of B_0 (all other parameters are those given above), has been plotted while using the forward (pink) and backward (cyan) methods. The corresponding Lyapunov exponents for the same range of parameter B_0 are reported in Figures 2.1b–2.1c.

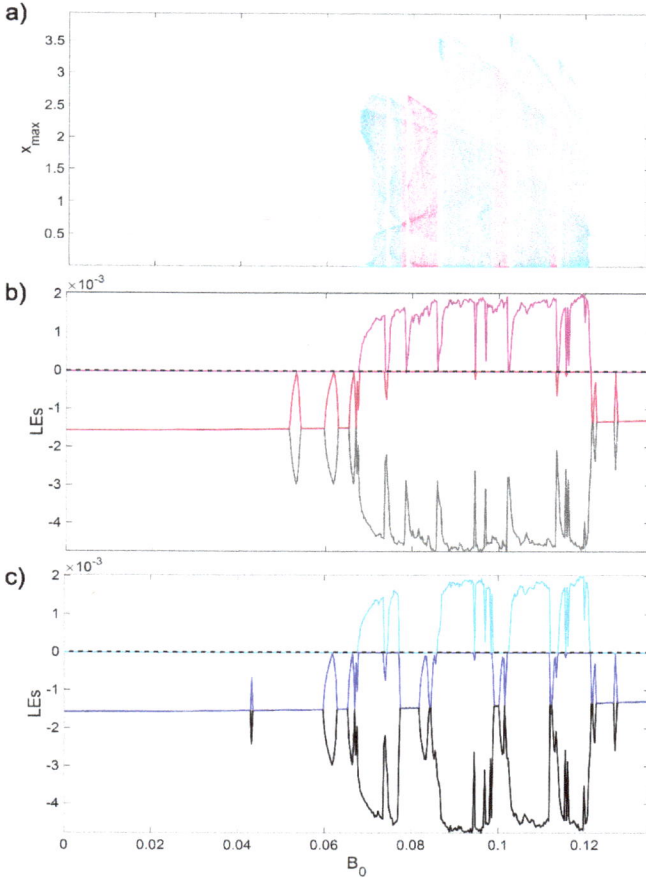

Figure 2.1. Bifurcation and Lyapunov exponents diagram as a function of B_0.

As we deduce from the bifurcation diagram (Figure 2.1a), the attractors' structure is rather complicated due to the presence of local and global bifurcations [Sprott (2003); Ott (1993)]. However, a qualitative description can be provided in terms of five leading periodic orbits present in this dynamical system (2.2). Such periodic solutions are labeled P_1, P_2, ... P_5, each of them is characterized by the presence of a single peak every 1 to 5 periods of the driving frequency (f_{mod}). As the control parameter B_0 is increased, we observe that the P_2 solution replaces the P_1 solution due to a crisis. This solution (upper branch solution) coexists in the range $B_0 \in [0.0445, 0.0558]$ with the lower branch solution and its first subharmonic $P_{1,2}$ (first bistable window B_I). The upper branch solution P_2 loses its stability via a subharmonic bifurcation till a new crisis is encountered and replaced by a P_3 solution which coexists with a lower branch chaotic attractor in the range $B_0 \in [0.0776, 0.0858]$ (this is the second bistability region). This process is continued up to the appearance of the P_4 solution and the fourth bistability region around $B_0 \in [0.0988, 0.1018]$. The fourth bistability occurs at around $B_0 \in [0.1120, 0.1134]$. At $B_0 \approx 0.12$, we observe the last crisis with lower amplitude attractors belonging to an inverse cascade of the primary P_1 lower branch solution. The four bistable regions $B_{I,II,III,IV}$ have been identified by using different numerical algorithms as the bifurcation parameter is scanned in the forward and backward directions [Jafari *et al.* (2021)]. Relevant information about the organization of the solutions is provided by the evaluation of the dissipation rate (trace of the Jacobian matrix). The corresponding dissipation diagram of the bifurcation diagram described above is shown in Figure 2.2.

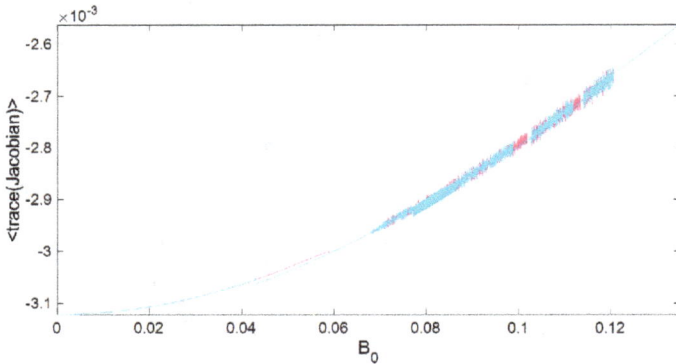

Figure 2.2.   Dissipation diagram $Tr(J)$ as a function of B_0.

From Figure 2.2, we can graphically deduce the bistability regions as well as in Figure 2.1 (see the zoomed regions of bistability in Figure 2.3). We observe that in the second bistability region B_{II} the upper branch periodic solution P_3 is characterized by the alternating between two values whose average is below the competing chaotic attractor. Figures 2.2 and 2.3 enable us to distinguish the two coexisting solutions according to their dissipative rate. It is important to note that the upper branch solutions (periodic solutions) are characterized by lower dissipativity.

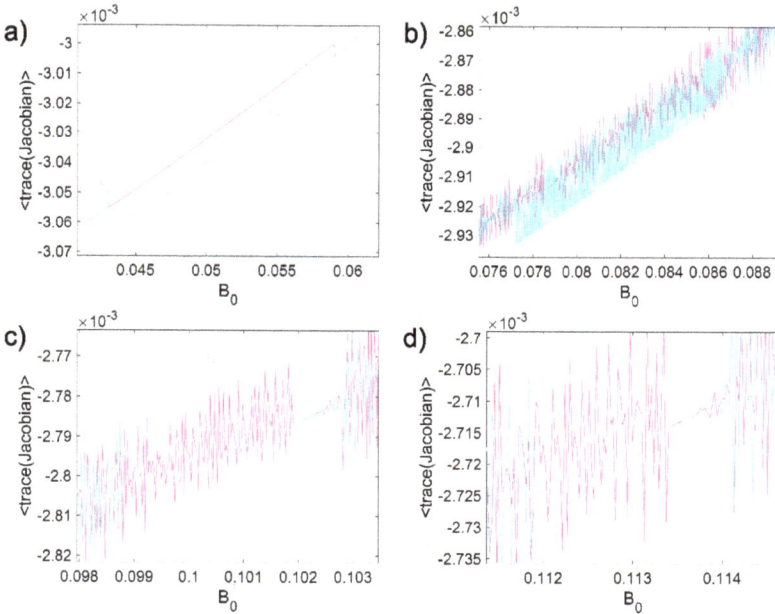

Figure 2.3.   Zoom in of bistability regions in Figure 2.2.

In Figure 2.4, the bifurcation diagram of the system (2.2), i.e. the maxima of the solution $x(t)$ as a function of γ as a function of gamma (all other parameters are those given above), has been plotted. From Figure 2.4 we observe three bistability windows around $\gamma = 2.5 \times 10^{-3}$, $\gamma = 3.5 \times 10^{-3}$ and $\gamma = 5.5 \times 10^{-3}$. In the latter bistability window, the two coexisting solutions are periodic ones. From the bifurcation diagram (Figure 2.4), it is also possible to identify the bifurcation regions.

Thus, in Section 2.2, we have numerically shown that generalized bistability (simultaneous presence of two kinds of attractors having the same

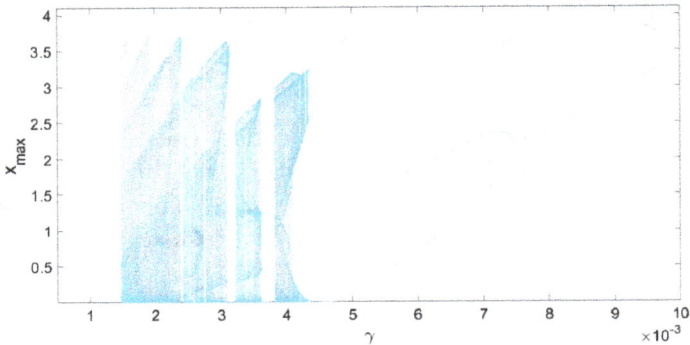

Figure 2.4.   Bifurcation diagram $x^{max}(t)$ as a function of γ.

values of the control parameters but different initial conditions, in our case, we only consider $x(0)$ is different) depends on either B_0 or γ for the two-level laser model (2.2). Such bistability regions will be highlighted with the help of a new kind of *bifurcation diagrams* presented in the next section below. To this aim, in the following section, we will transform this two-level non-autonomous model (2.2) into a four-dimensional autonomous dynamical system.

2.3   Two-level autonomous laser model

Let us notice that the presence of the $\sin(2\pi f_{mod}t)$ on the right-hand side of the first equation of the two-level model (2.2) makes it non-autonomous. However, recall that a sine function is nothing else but the solution of a harmonic oscillator. So, let us take:

$$z(t) = B_0 + m\sin(2\pi f_{mod}t),$$

which is the solution of the following second-order ordinary differential equation (ODE):

$$\ddot{z}(t) + \omega^2 z(t) = \omega^2 B_0.$$

where $\omega = 2\pi f_{mod}$. By using the classical D'Alembert transformation [D'Alembert (1748)], this second-order ODE may be written as the following system of first-order ODEs:

$$\dot{z} = -\omega^2 u(t), z(0) = B_0,$$
$$\dot{u} = z(t) - B_0, u(0) = -\frac{m}{\omega}.$$

Thus, we transform the two-level non-autonomous model (2.2) into an autonomous one while increasing the dimension of two. We have:

$$
\begin{aligned}
\dot{x} &= -x\left(1 + kz^2 - y\right), \\
\dot{y} &= -\gamma y - \alpha xy + \gamma p_0, \\
\dot{z} &= -\omega^2 u, \\
\dot{u} &= z - B_0,
\end{aligned}
\tag{2.5}
$$

where $k = 12$, $B_0 = 0.1215$, $m = 0.02$, $f_{mod} = 0.005$, $\omega = 2\pi f_{mod}$, $\gamma = 0.0025$, $\alpha = 0.002$, $p_0 = 1.252$ and where the initial conditions $z(0) = B_0$, $u(0) = -m/\omega$ are imposed on $z(t)$ and $u(t)$.

2.3.1 Fixed points

By using the classical nullclines method, it can be shown that the dynamical system (2.5) admits two fixed points.

$$I_1\left(0, p_0, B_0, 0\right) \; ; \; I_2\left(-\frac{\gamma}{\alpha}\frac{1 + kB_0^2 - p_0}{1 + kB_0^2}, 1 + kB_0^2, 0, 0\right) \tag{2.6}$$

2.3.2 Jacobian matrix

The Jacobian matrix of the dynamical system (2.5) reads:

$$
J =
\begin{pmatrix}
-(1 + kz^2 - y) & x & -2kxz & 0 \\
-\alpha y & -\gamma - \alpha x & 0 & 0 \\
0 & 0 & 0 & -\omega^2 \\
0 & 0 & 1 & 0
\end{pmatrix}
\tag{2.7}
$$

By replacing the coordinate of the fixed points I_1 (2.6) in the Jacobian matrix (2.7), one obtains the four following eigenvalues:

$$\lambda_1 = -\gamma, \; \lambda_2 = -\left(1 + kB_0^2 - p_0\right), \; \lambda_{3,4} = \pm i\omega. \tag{2.8}$$

With this parameters set $-(1 + kB_0^2 - p_0) > 0$, λ_2 is real and positive while λ_1 is real and negative. So, according to the Lyapunov theorem, the fixed point I_1 is *unstable*.

By replacing the coordinate of the fixed points I_2 (2.6) in the Jacobian matrix (2.7), one obtains the four following eigenvalues:

$$\lambda_{1,2} = -\frac{\gamma p_0}{2y^*} \pm \frac{\sqrt{\Delta}}{2} \quad \text{and} \quad \lambda_{3,4} = \pm i\omega. \tag{2.9}$$

where

$$\Delta = (\frac{\gamma p_0}{y^*})^2 + 4\gamma(y^* - p_0) \quad \text{with} \quad y^* = 1 + kB_0^2.$$

With this set of parameters, $\Delta < 0$ and both real parts gamma $\gamma p_0/y^*$ of $\lambda_{1,2}$ are negative. Hence, according to the Lyapunov theorem, the fixed point I_2 is *stable*. Notice that a Hopf bifurcation could only occur if and only if $p_0 = 0$. Nevertheless, in such a case, Δ would become positive. So, no Hopf bifurcation can occur in system (2.5). It is important to note that the emergence of chaos is related to the interplay between the two stationary points. The unstable fixed point I_1 provides the re-injection to I_2, allowing time evolution of the trajectory in phase space.

Now, let us highlight the bistability regions of the system (2.5) and so of the two-level model (2.2). To this aim, we propose to use *bistability bifurcation diagrams* by plotting (as usual) the maxima of the variable $x(t)$ as a function of the initial condition $x(0)$ instead of the control parameter B_0 which is fixed here. Figures 2.5a–2.5d, represent such bifurcation diagrams for $B_0 = 0.05$, $B_0 = 0.08$, $B_0 = 0.1$ and $B_0 = 0.112$.

Figure 2.5 highlights the existence of two different regions of stability for the attractor solution of the system (2.5). In the following, we will consider that the lower branch corresponds to the first stability region while the upper branch corresponds to the second. As an example, from Figure 2.5a ($B_0 = 0.05$), we deduce that for $x(0) = 1$, the attractor is in the lower branch, and the solution is a periodic solution with one peak for the $x(t)$. For $x(0) = 3$, the attractor is in the upper branch, and the solution is a P_2 solution (see Figure 2.6a). From Figure 2.5b ($B_0 = 0.08$), we find that for $x(0) = 1$, the attractor is in the lower branch and the solution is chaotic. For $x(0) = 3$, the attractor is in the upper branch, and the periodic solution is a P_3 solution (see Figure 2.6b). From Figure 2.5c ($B_0 = 0.1$), we find that for $x(0) = 1$, the attractor is in the lower branch and the solution is

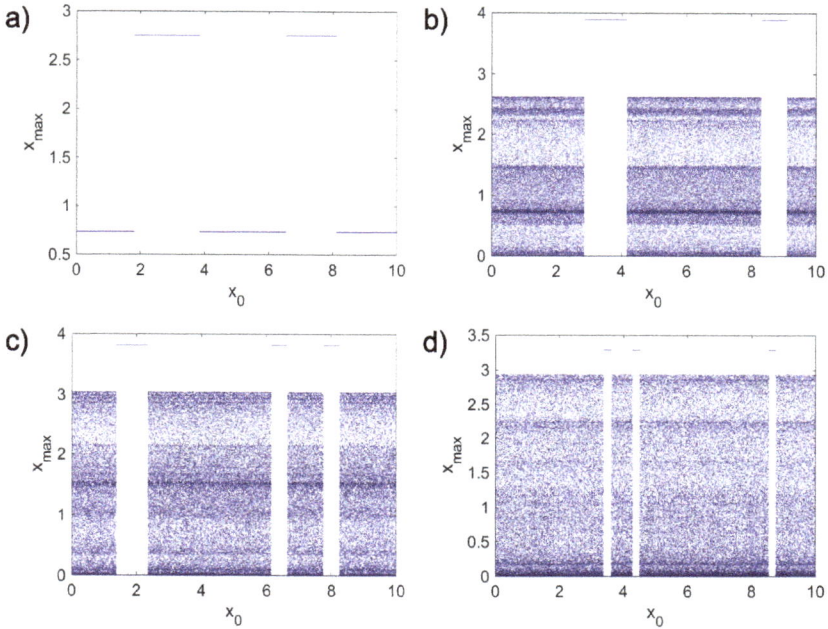

Figure 2.5. Bistability diagrams of system (2.5). Maxima of the solution $x(t)$ as a function of the initial condition $x(0)$ for four different values B_0 corresponding to the four bistability windows.

chaotic. For $x(0) = 2$, the attractor is in the upper branch, and the periodic solution is a P_4 (see Figure 2.6c). From Figure 2.5d ($B_0 = 0.112$), we find that for $x(0) = 3.7$, the attractor is in the lower branch and the periodic solution is a P_5. For $x(0) = 4$, the attractor is in the upper branch, and the solution is chaotic (see Figure 2.6d). Notice that both attractors coexist in all these cases. The basins of attraction corresponding to the four bistability regions are reported in Figure 2.7. Although basins of attraction provide general information about the organization of the phase space, we consider the representation in terms of bistability bifurcation diagrams easier to be interpreted.

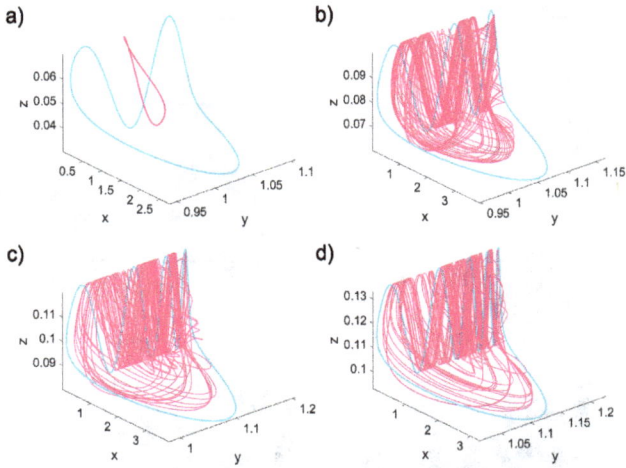

Figure 2.6.   Phase portraits of system (2.5) in the xyz-space for various values B_0.

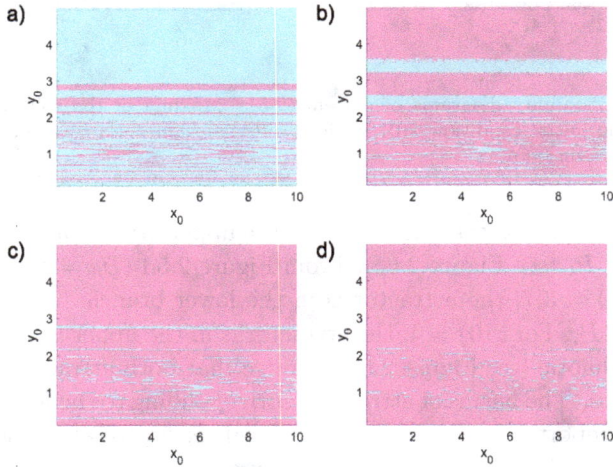

Figure 2.7.   Basins of attraction of system (2.5) for various values of B_0 as in Figure 2.6.

2.4   Control of generalized multistability

Let us consider the effects of a second sinusoidal perturbation with amplitude ε smaller than m in the two-level non-autonomous model (2.2). So, the dynamics are now described by:

$$\dot{x} = -x\left[1 + k\left(B_0 + \varepsilon\sin(2\pi f_{mod}t + \varphi) + m\sin(2\pi f_{mod}t)\right)^2 - y\right],$$
$$\dot{y} = -\gamma y - \alpha xy + \gamma p_0.$$
(2.10)

where $k = 12$, $B_0 = 0.0832$, $m = 0.02$, $\varepsilon = 0.0006$, $f_{mod} = 0.005$, $\gamma = 0.0025$, $\alpha = 0.002$ and $p_0 = 1.252$. Bistability can be removed by accurately choosing ε and phase difference φ as we can observe from Figure 2.8. In such a case the optimal value for the phase difference φ is around $\pi = 3.1415$. The opposite effect occurs when $\phi = 0 = 2\pi$ as we can observe from Figure 2.9. Removing bistability means to stabilize to lower amplitude attractors which are less dissipative compared with the upper branch solutions (see the blue and pink traces in Figure 2.8).

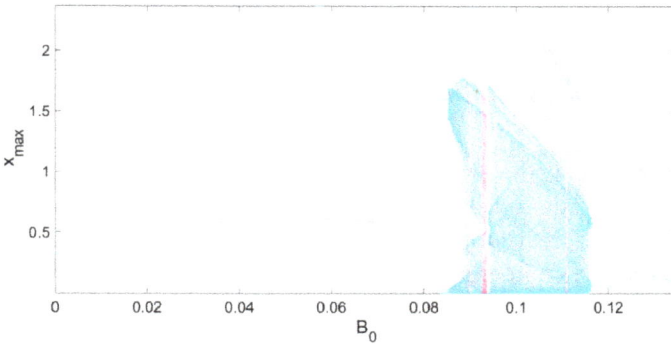

Figure 2.8.   Bifurcation diagram $x^{max}(t)$ as a function of B_0 with $(\varepsilon, \varphi) = (0.006, \pi)$.

From a comparison with the unperturbed bifurcation diagram shown in Figure 2.1, we see that the controlling perturbation critically affects the dynamics by delaying the occurrence of the first bifurcation window to a value of the control parameter B_0, where the unperturbed dynamics was chaotic. Furthermore, we note that, for $\varepsilon = 0.006$, which implies a relative perturbation strength $\varepsilon/m = 0.006/0.02 = 30\%$, the unperturbed bistability window centered around $B_0 = 0.05$ is shifted around $B_0 = 0.06$. Here several considerations can be drawn. First, if the competing solutions are periodic, as in the first bistability window, we need to use a greater perturbation with respect to the case where the lower branch solution is chaotic, and the upper branch is periodic. From Figure 2.8, we clearly see that the first bistability window is shifted in the forward B_0 direction, but the other bistability windows are removed except the small one around $B_0 = 0.09$. This is consistent with the fact that the width of the bistability regions diminishes as B_0 is increased.

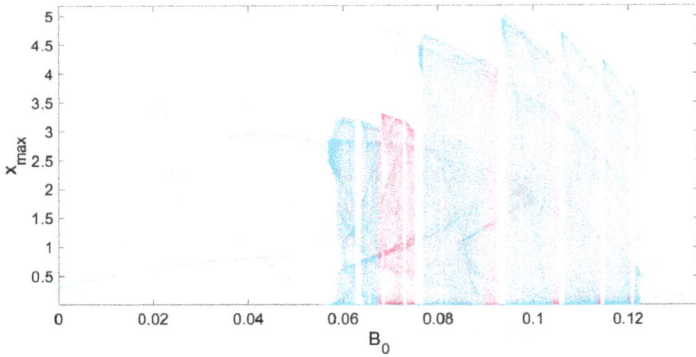

Figure 2.9.   Bifurcation diagram $x^{max}(t)$ as a function of B_0 with $(\varepsilon, \varphi) = (0.006, 0)$.

In Figure 2.9, where the wrong phase difference φ is chosen, we clearly see that the first bistability window and the other ones are anticipated with respect to the bifurcation parameter B_0. We also observe that their width is increased when compared with the unperturbed case of Figure 2.1.

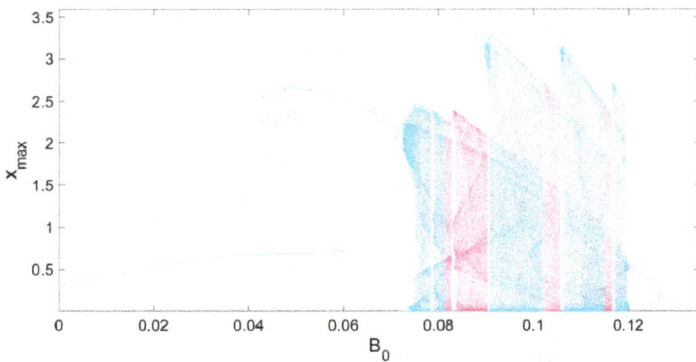

Figure 2.10.   Bifurcation diagram $x^{max}(t)$ as a function of B_0 with $(\varepsilon, \varphi) = (0.002, \pi)$.

As the perturbation strength ε is reduced (keeping the right phase difference φ), we observe that the adopted strategy for controlling bistability remains valid but on restricted regions of the unperturbed bistable regions. This fact emerges from Figure 2.10 where the relative perturbation strength is $\varepsilon/m = 0.002/0.02 = 10\%$.

2.5 Discussion

The original laser model that gave evidence of chaos and generalized multistability has been revisited. Multistability depends on dissipativity, so the two decay rates for laser intensity and population inversion must be chosen accurately. Multistability can be controlled by means of a secondary sinusoidal perturbation resonant with the main one responsible for its occurrence. The stabilized solution is the one characterized by less dissipativity providing an important indicator for applications when the selection of the solutions is desired.

2.6 References

Arecchi F.T., Meucci R., Puccioni G.P. & Tredicce J.R. (1982). Experimental evidence of subharmonic bifurcations, multistability, and turbulence in a Q-switched gas laser, *Physical Review Letters*, 49(17), pp. 1217-1220.

Feudel U. (2008). Complex dynamics in multistable systems, *Int. J. Bifurcation Chaos*, 18(6), pp. 1607-1626.

Feudel U., Pisarchik A.N. & Showalter K. (2018). Multistability and tipping: From Mathematics and physics to climate and brain — Minireview and preface to the focus issue, *Chaos*, 28, 033501 (8 pages).

Saucedo-Solorio J.M., Pisarchik A.N., Kir'yanov A.V. & Aboites V. (2003). Generalized multistability in a fiber laser with modulated losses, *J. Opt. Soc. Am. B*, 20, pp. 490-496.

Pisarchik A.N., Jaimes-Reategui R., Sevilla-Escoboza R., Huerta-Cuellar G. & Taki M. (2011). Rogue waves in a multistable system, *Phys. Rev. Lett.*, 107, 274101 (5 pages).

Schwartz J.-L., Rimault N.G., Hupe J.-M., Moore B. & Pressnitzer D. (2012). Multistability in perception: Binding sensory modalities, an overview, *Philos. Trans. R. Soc. B*, 367, pp. 896-905.

Ryashko L. (2018). Sensitivity analysis of the noise-induced oscillatory multistability in Higgins model of glycolysis, *Chaos*, 28, 033602.

Lucarini V. & Bodai T. (2017). Edge states in the climate system: Exploring global instabilities and critical transitions, *Nonlinearity*, 30, R32-R66.

May R. (1977). Thresholds and breakpoints in ecosystems with multiplicity of stable states, *Nature*, 269, pp. 471-477.

Ciofini M., Politi A. & Meucci R. (1993). Effective two-dimensional model for CO_2 lasers, *Phys. Rev. A*, 48, pp. 605-610.

Meucci R., Euzzor S., Arecchi F.T. & Ginoux J.-M. (2021). Minimal universal model for chaos in a laser with feedback, *Int. J. Bifurcation Chaos*, 31(4), 2130013 (10 pages).

Ricci L., Perinelli A., Castelluzzo M., Euzzor S. & Meucci R. (2021). Experimental evidence of chaos generated by a minimal universal oscillator model, *Int. J. Bifurcation Chaos*, 31(12), 2150205 (13 pages).

Goswami B.K. & Pisarchick A.N. (2008). Controlling multistability by small periodic perturbation, *Int. J. Bifurcation Chaos*, 18(6), pp. 1645-1673.

Meucci R., Gadomski W., Ciofini M. & Arecchi F.T. (1994). Experimental control of chaos by means of weak parametric perturbations, *Physical Review E*, 49(4), R2528-R2531.

Meucci R., Euzzor S., Pugliese E., Zambrano S., Gallas M.R. & Gallas J.A.C. (2016). Optimal phase-control strategy for damped-driven Duffing oscillators, *Physical Review Letters*, 116(4), 044101 (5 pages).

Arecchi F.T., Gadomski W. & Meucci R. (1986). Generation of chaotic dynamics by feedback on a laser, *Phys. Rev. A*, 34(2), pp. 1617-1620.

Arecchi F.T., Meucci R. & Gadomski W. (1987). Laser dynamics with competing instabilities, *Phys. Rev. Lett.*, 58(21), pp. 2205-2208.

Grebogi C., Ott E. & Yorke J.A. (1983). Crises: Sudden changes in chaotic attractors and chaotic transients, *Physica D*, 7, pp. 181-200.

Feudel U., Grebogi C., Hunt B.R. & Yorke J.A. (1996). Map with more than 100 coexisting low-period periodic attractors, *Phys. Rev. E*, 54, pp. 71-81.

Meucci R., Poggi A., Arecchi F.T. & Tredicce J.R. (1988). Dissipativity of an optical chaotic system characterized via generalized multistability, *Opt. Commun.*, 65, pp. 151-156.

Sprott J.C. (2003). *Chaos and Time-Series Analysis* (Oxford University Press).

Ott E. (1993). *Chaos in Dynamical Systems* (Cambridge University Press).

Jafari A., Hussain I., Nazarimehr F., Golpayegani S.M.R.H. & Jafari S. (2021). A simple guide for plotting a proper bifurcation diagram, *Int. J. Bifurcation Chaos*, 31(1), 2150011 (11 pages).

D'Alembert J. (1748). Suite des recherches sur le calcul intégral, quatrième partie : Méthodes pour intégrer quelques équations différentielles, *Hist. Acad. Berlin*, tome IV, pp. 275-291.

Chapter 3

Minimal Universal Model for Chaos in Laser with Feedback

In this chapter, we revisit the model of the laser with feedback and the minimal nonlinearity leading to chaos. Although the model has its origin in laser physics, with peculiarities related to the CO_2 laser, it belongs to the class of the three-dimensional paradigmatic nonlinear oscillator models generating chaos. The proposed model contains three key nonlinearities, two of which are of the type xy, where x and y are the fast and slow variables. The third one is of the type xz^2, where z is an intermediate feedback variable. We analytically demonstrate that it is essential for producing chaos via local or global homoclinic bifurcations. Its electronic implementation in the range of kilo Hertz (kHz) region confirms its potential in describing phenomena evolving on different time scales.

3.1   Introduction

Deterministic chaos has represented a crucial issue in laser physics because the Lorenz system [1963] is formally equivalent to laser equations [Haken (1975)]. The first evidence of chaos in lasers was given in 1982 in a modulated single mode CO_2 laser [Arecchi *et al.* (1982)] and we had to wait until 1986 for an evidence of Lorenz type chaos in a particular and little used laser emitting in far infrared region (the so-called class C-laser) [Weiss & Brock (1986)]. Commonly used single mode lasers are described by two rate equations and they are intrinsic stable devices (class B-laser). Feedback circuits are frequently used to improve their stability properties by reducing residual intensity and frequency fluctuations in order to match specific requests. However, a feedback can have the opposite effect, that is, enhancing the relaxation oscillations around the steady state solution. In other words, a simple linear filtering with the appropriate bandwidth on

33

the laser output intensity can induce chaotic fluctuations on it [Arecchi *et al.* (1986); Arecchi (1987)]. From the dynamical point of view, a feedback increases the dimensionality of the phase space from two to three whence chaos becomes possible. In the last two decades of the past century, the use of CO_2 lasers for demonstrating such a behavior has had advantages, mainly due the convenient time scales associated with the laser intensity and the population inversion. The former, evolves on a time scale regulated by the parameter k (the decay rate of photon number of the laser mode) is imposed by the length of the optical cavity and its losses. In our case, k is around $10^7 \ s^{-1}$. The latter, i.e. the decay rate of the population inversion, is called γ. For a molecular laser as the CO_2 laser, γ is of the order $10^3 - 10^4$ s^{-1}. These two parameters imply a resulting time scale given by

$$\sqrt{k\gamma(p_0 - 1)},$$

where p_0 is the pump strength normalized to the threshold value. Usually, p_0 is around 2. To be effective in producing chaos, the feedback loop should act on the above mentioned time scale, in other terms, we have the following condition to be satisfied:

$$k > \beta > \gamma,$$

where β is the bandwidth of the feedback loop. A simple three-dimensional model accounting for the laser intensity x with decay rate k, population inversion y with decay rate γ, and feedback strength z with decay rate β, explains qualitatively the observed dynamics but it does not yield an accurate matching with the experiment. The problem is overcome by introducing the so-called 4-level model for the CO_2 laser. It consists in taking into account two resonant levels with populations N_1 and N_2 and two rotational manifolds with populations M_1 and M_2, respectively. In this refined model the dimensionality of the phase space is increased up to 6 instead of the previous 3. For an introduction to the six-dimensional model and accurate numerical simulations on it, see for example Freire *et al.* [2015]. To reduce the dimensionality two different approaches can be followed. The first one is to use a reduction based on the Center Manifold Theory (CMT) proposed by Varone *et al.* [1995]. This analytical method implies a reduced four-dimensional model with the addition of nonlinear terms whose physical interpretation is difficult to provide. A feasible reduction to three dimensions is imaginable considering that it has been obtained for the five-dimensional model of the CO_2 laser with cavity losses modulation [Ciofini *et al.* (1993)]. The second approach for an equivalent three-dimensional model which is more physical and straightforward is to take advantage of the 4-level model only from the correct value of the laser intensity in the stationary regime. This condition implies the use of an artificial value of γ which can be of two orders of magnitude greater, that is, 10^4 to 10^5 s^{-1}. As the laser output intensity does not depend on β, we have to use an effective value of β for the feedback variable z up to 10^6 s^{-1}. The advantage consists in keeping the original nonlinearity in the x and y differential equations given by their product xy. Considering the above advantages and in view of the extension to different dynamical systems ranging from neuron dynamics in the low frequency region (below 1 Hz) to high frequency domains (fast electronics, opto-electronics, etc.) we adopt the following three-dimensional model:

$$\frac{dx}{dt} = -k_0 x \left(1 + k_1 z^2 - y\right),$$

$$\frac{dy}{dt} = -\gamma y - 2\frac{k_0}{\alpha} xy + \gamma p_0, \tag{3.1}$$

$$\frac{dz}{dt} = -\beta \left(z - B_0 + \frac{R}{\alpha} x\right),$$

where x is the fast variable (laser output intensity), y is the slow variable (population inversion), and z is the feedback variable affecting the fast one in a nonlinear way but regulated in linear way as the result of a low pass filter whose input is the fast variable summed to bias. The parameter α is a suitable normalization of the fast variable x which however does not alter its form considering that it is homogenous in x. If $\alpha = 2k_0/\gamma$, the adopted model is formally equivalent to the original model of the laser with feedback. The chapter is organized as follows. First, we introduce a numerical analysis followed by an experimental part containing an analog implementation of the oscillator. Second, we compare the new model with other paradigmatic oscillators. Its potentialities are discussed in conclusions.

3.2   Minimal universal model

3.2.1   *Dimensionless form*

We propose the following change of variables and parameters to recast Eq. (3.1) in a dimensionless form. Let us take:

$$y \to p_0 y, \; t \to \frac{t}{\gamma}, \; \epsilon_1 = \frac{k_0}{\gamma}, \; \epsilon_2 = \frac{\beta}{\gamma}, \; B_1 = \frac{R}{\alpha} = \frac{R\gamma}{2k_0}.$$

Thus, the minimal universal model for chaos in laser reads:

$$\frac{dx}{dt} = -\epsilon_1 x \left(1 + k_1 z^2 - p_0 y\right),$$

$$\frac{dy}{dt} = -y - xy + 1, \tag{3.2}$$

$$\frac{dz}{dt} = -\epsilon_2 \left(z - B_0 + B_1 x\right),$$

Let us note that with the parameter set used in our experiment and analysis, $\epsilon_1 \gg 1$ and $\epsilon_2 \gg 1$. So, model (3.2) is a *slow-fast* dynamical systems involving two *fast* times scales. In the following, B_0 will play the role of a control parameter.

3.2.2   *Fixed points*

By using the classical nullclines method, it can be shown that the dynamical system (3.2) admits four fixed points only two of which are positive.

$$I_1 \left(0, 1, B_0\right),$$

$$I_2 \left(x^*, y^* = \frac{1}{1+x}, z^* = B_0 - B_1 x\right), \tag{3.3}$$

where the expression of x^* (too large to be explicitly written here since it comes from the solution of a cubic polynomial) depends on the control parameter B_0. In this problem all fixed points are supposed to be positive. Thus, starting from the right-hand-side of Eq. (3.2), it can be easily shown that:

$$0 \leqslant x^* \leqslant p_0 - 1, \; \frac{1}{p_0} \leqslant y^*, \; B_0 - B_1 \left(p_0 - 1\right) \leqslant z^* \leqslant B_0.$$

3.2.3 *Jacobian matrix*

The Jacobian matrix of dynamical system (3.2) reads:

$$
J = \begin{pmatrix}
(-1 + p_0 y - k_1 z^2)\epsilon_1 & p_0 x \epsilon_1 & -2k_1 xz\epsilon_1 \\
-y & -1 - x & 0 \\
-B_1 \epsilon_2 & 0 & \epsilon_2
\end{pmatrix}
\tag{3.4}
$$

By replacing the coordinate of the fixed points I_1 (3.3) in the Jacobian matrix (3.4) one obtains the Cayley–Hamilton third degree eigenpolynomial which reads:

$$
(\lambda + 1)(\lambda + \epsilon_2) \left[\lambda + \epsilon_1 \left(1 + B_0^2 k_1 - p_0 \right) \right] = 0
\tag{3.5}
$$

Thus, provided that:

$$
p_0 < 1 + B_0^2 k_1,
\tag{3.6}
$$

the fixed point I_1 is a *saddle node*. Moreover, such condition (3.6) provides a upper boundary for the control parameter B_0:

$$
B_0 < \sqrt{\frac{p_0 - 1}{k_1}}.
\tag{3.7}
$$

From the positivity of the fixed points, we notice that if $B_0 = B_1(p_0 - 1)$, then the second fixed point I_2 reads: $I_2(x^* = p_0 - 1, y^* = 1/p_0, z^* = 0)$. In these conditions and according to Eq. (3.7), it can be stated that:

$$
B_1(p_0 - 1) < B_0 < \sqrt{\frac{p_0 - 1}{k_1}}.
\tag{3.8}
$$

Thus, for $B_0 = B_1(p_0 - 1)$, computation of the eigenvalues of I_1 shows that two of them are real and negative and one is real and positive confirming thus the *saddle node* feature of this point while for I_2, one is real and negative and the two others are complex conjugate with negative real parts. So, in this case the fixed point I_2 is stable and attractive according to Lyapunov theorem. For $B_0 = \sqrt{(p_0 - 1)/k_1}$, we found that both fixed points I_1 and I_2 are stable and attractive (all the real parts of their eigenvalues are negative). Such a result will enable to explain the limits of the bifurcation diagram presented below (see Figures 3.1 and 3.2) outside which no attractor can exist. Then, while using the parameters set of our experiment, i.e. for any value of $B_0 \in [B_1(p_0 - 1), \sqrt{(p_0 - 1)/k_1}]$ it can be shown that I_2 is a *saddle-focus* (the first eigenvalue is real and negative while the two others are complex conjugate with positive real parts). This implies that a Hopf bifurcation occurs in the interval [Hopf (1942); Andronov *et al.* (1971); Marsden & McCracken (1976); Kuznetsov (2004)]. Nevertheless, the fact that the fixed points are solutions of a cubic polynomial precludes from the explicit value of the control parameter B_0 for which such a Hopf bifurcation occurs. However, we have found a method (see Appendix) allowing to analytically compute the upper bound of such a parameter. In the following we will use the parameters set in our experiment and analysis:

$$\epsilon_1 = 200 \ , \ \epsilon_2 = 6 \ , \ k_1 = 12 \ , \ p_0 = 1.208 \ , \ B_1 = 0.555.$$

3.2.4   *Bifurcation diagram*

Thus, in order to highlight the effects of the control parameter B_0 changes on the topology of the attractor, we have built a bifurcation diagram (see Figures 3.1 and 3.2) that we have compared to the phase portraits plotted in Figure 3.3. First, we observe that for $B_0 \approx 0.12$, a Hopf bifurcation occurs (see Appendix). Then, for $B_0 = 0.123$, a *limit cycle* appears. As B_0 increases between 0.123 and 0.1237, a "period doubling cascade" occurs and so, we observe a *2-periodic limit cycle*. In the interval $0.1237 < B_0 < 0.12425$, the period of the *limit cycle* increases again and becomes equal to the number of branches in the bifurcation diagram (see Figures 3.2 and 3.3a). For $0.12425 < B_0 < 0.129$, a *stable homoclinic orbit* appears and persists (see Figures 3.2d and 3.3b).

In order to confirm such scenario, Lyapunov Characteristic Exponents (LCE) have been computed in each case.

3.2.5   *Numerical computation of the Lyapunov exponents*

The algorithm developed by Marco Sandri [1996] for Mathematica® has been used to perform the numerical calculation of the Lyapunov Characteristics Exponents (LCE) of the dynamical system (3.2) in each case. LCEs values have been computed within each considered interval ($B_0 \in [0.123, 0.1234]$ and $[0.1235, 0.125]$). As an example, for $B_0 = 0.123$, 0.124 and 0.1246, Sandri's algorithm has provided respectively the following LCEs $(0, -0.56, -6.63)$, $(+0.27, 0, -7.47)$ and $(+0.2, 0, -7.32)$.

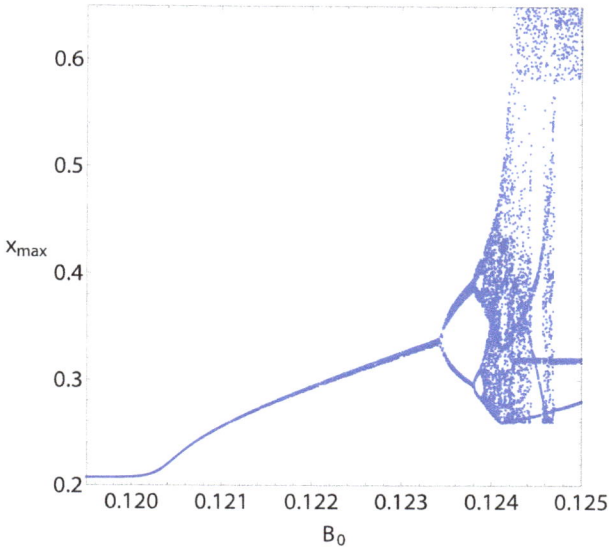

Figure 3.1.   Bifurcation diagram x_{max} as a function of B_0.

Then, following the works of Klein and Baier [1991], a classification of (autonomous) continuous-time attractors of dynamical system (3.2) on the basis of their Lyapunov spectrum is presented in Table 3.1. LCEs values have also been computed with the Lyapunov Exponents Toolbox (LET) developed by Steve Siu for MatLab® and involving the two algorithms proposed by Wolf *et al.* [1985] and Eckmann and Ruelle [1985] (see https://fr.mathworks.com/matlabcentral/fileexchange/233-let).   Results obtained by both algorithms are consistent.

Table 3.1.   Lyapunov characteristic exponents of dynamical system (3.2) for various values of B_0.

m	LCE spectrum	Dynamics of the attractor
$0.1230 < B_0 < 0.1234$	$(0, -, -)$	Periodic Motion
$0.1235 < B_0 < 0.124$	$(0, -, -)$	n-Periodic Motion
$0.1241 < B_0 < 0.125$	$(+, 0, -)$	Homolinic Chaos

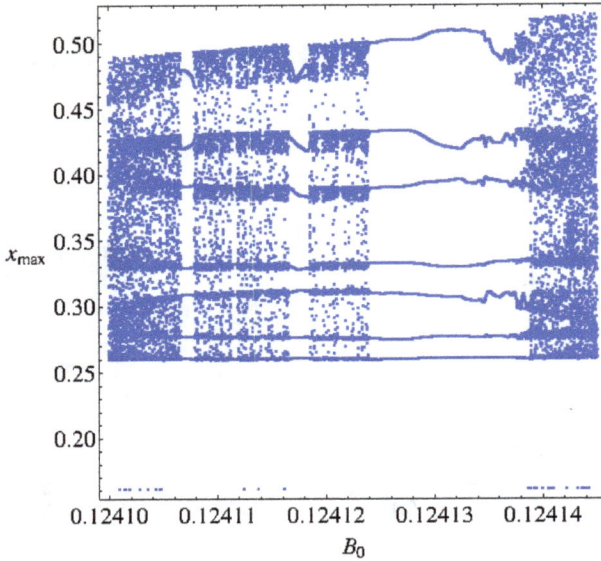

Figure 3.2.   Zoom of the bifurcation diagram x_{max} as a function of B_0.

3.3   Experimental part

The set up used in our experiment is illustrated in Figure 3.7. Briefly, it consists of three integrators I_1, I_2 and I_3 (LT1114 by Analog Devices), whose outputs are the signals x, y and z and $y = \dot{x}$ contained in Eq. (3.1). The other integrator I_4, is employed for an inverting amplifier with unitary gain. The two nonlinearities are implemented by means of three analog multipliers M_1, M_2 and M_3 (MLT04, by Analog Devices). The first one yields the product xy, while the other two multipliers implement the product xz^2. The simplification of the proposed scheme is evident when compared with the one obtained using a Field Programmable Analog Array (FPAA) circuit [Arecchi *et al.* (2005)].

Considering the limits imposed by analog simulations, the desired dynamics of the oscillator is obtained by fine adjustments of two bias voltages $V(p_0)$ and $V(B_0)$ accounting for the parameters, p_0 and B_0 in Eq. (3.1). This condition has been achieved by means of two potentiometers P_1 and P_2 connected to a fixed negative voltage source $-V_s$. The relaxation rates of the three variables (reciprocal of the integration times of the three

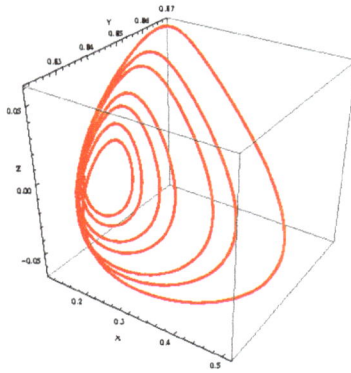

(a) $B_0 = 0.12413$

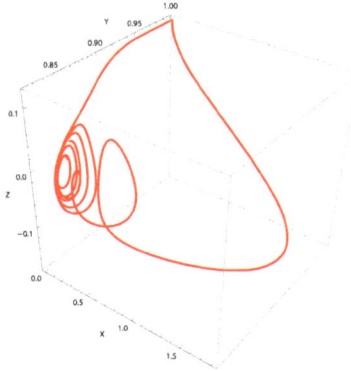

(b) $B_0 = 0.1243$

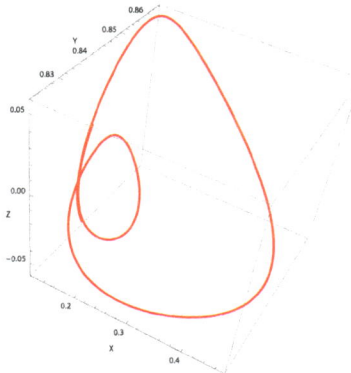

(c) $B_0 = 0.12463$

Figure 3.3. Phase portraits of model (3.2) for various values of B_0.

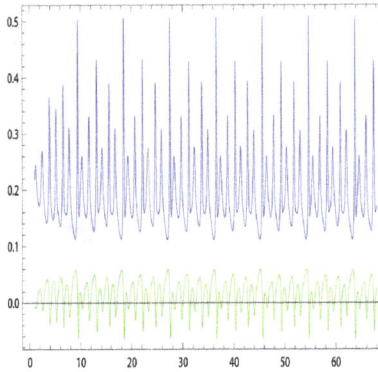

(a) Time series $x(t)$ and $z(t)$ for $B_0 = 0.12413$

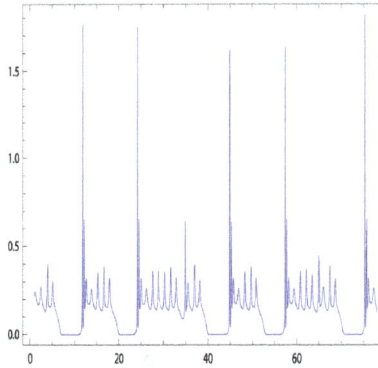

(b) Time series $x(t)$ for $B_0 = 0.1243$

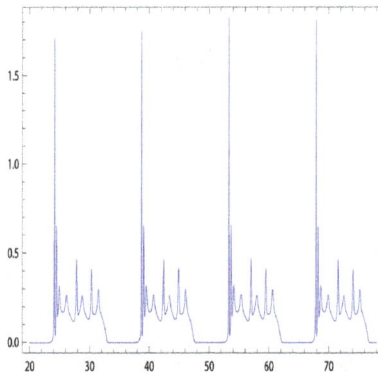

(c) Time series $x(t)$ for $B_0 = 0.12463$

Figure 3.4.   Time series of model (3.2) for various values of B_0.

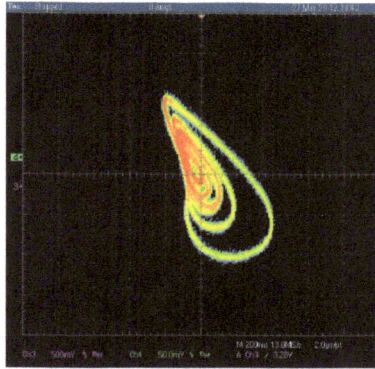

(a) Phase portrait for $V(B_0) = -2.550V$

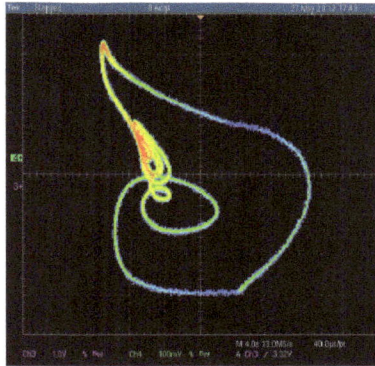

(b) Phase portrait for $V(B_0) = -2.596V$

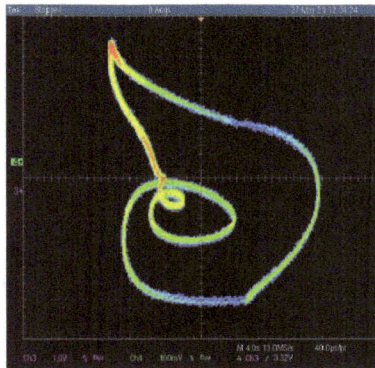

(c) Phase portrait for $V(B_0) = -2.628V$

Figure 3.5.   Oscilloscope snapshots of phase portraits for $V(p_0) = -2.983V$ and various values of $V(B_0)$.

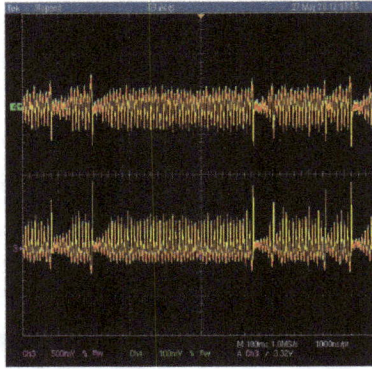

(a) Time series $x(t)$ and $z(t)$ for $V(B_0) = -2.550V$

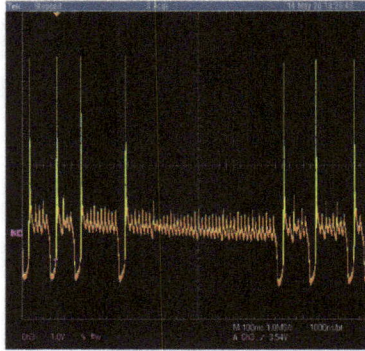

(b) Time series $x(t)$ for $V(B_0) = -2.596V$

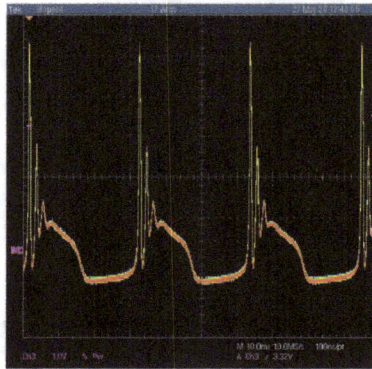

(c) Times series $x(t)$ for $V(B_0) = -2.628V$

Figure 3.6.   Oscilloscope snapshots of time series for $V(p_0) = -2.983V$ and various values of $V(B_0)$.

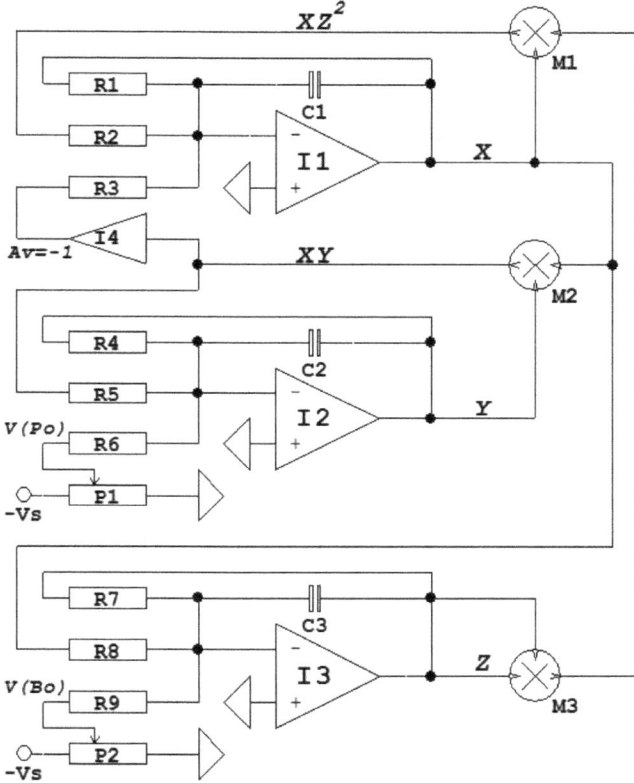

Figure 3.7. Circuit diagram of the "laser with feedback".

integrators), according to Eq. (3.1) and after a temporal rescaling of three orders of magnitude, are selected as follows. For the x integrator we set $k_0 = 1/(R_1C_1) = 2.463 \times 10^4 s^{-1}$ where $R_1 = 33k\Omega$ and $C_1 = 1.67nF$, $(R_3 = R_1 = 33k\Omega)$. For the y integrator we set $\gamma = 1/(R_4 \times C_2) = 1.00 \times 10^2 s^{-1}$, where $R_4 = 10k\Omega$, $C_2 = 1\mu F$. For the z integrator we select $\beta = 1/(R_7 \times C_3) = 1.00 \times 10^3 s^{-1}$, with $R_7 = 10k\Omega$ and $C_3 = 100nF$. For the other parameter values we have: $k_1 = R_1/(R_2 \times 2.5) = 13.2$ where $R_2 = 1k\Omega$; $p_0 = R_4/R_6 = 1.196$ where $R_6 = 8.36k\Omega$; $B_0 = R_7/R_9 = 0.115$ where $R_9 = 86.6k\Omega$; $R/\alpha = R_7/R_8 = 0.222$, where $R_8 = 45k\Omega$.

The attractors in the $x - z$ phase space for different values of the B_0 parameter value are reported in Figure 3.5. In panel (a) the local dynamics emerging after an Hopf bifurcation is shown. An increase of the control parameter B_0 leads to a condition of homoclinic chaos due to a homoclinic orbit around which a chaotic regime characterized by pulses nearly of the same height but erratically separated in time due to the rejection mechanism around the local chaos are shown in (a). A successive increment of the control parameter leads to the stabilization of the pulsed regime which are typical in a relaxation oscillator. In the presented attractors (see Figures 3.5b and 3.5c) we observe a small distortion for high amplitudes of the x signal due to the frequency limitation of the analog multipliers. This limitation does not induce severe limitations to the global dynamics as shown in the temporal behavior of the x signal, when we consider applications in the low frequency range below 1 kHz. For applications in high frequency regimes the nonlinearities must be implemented by using CMOS devices (range 1–10 MHz).

3.4   Discussion

Numerical analysis of the proposed model and its electronic implementation confirm its potentialities that promote it to the class of other well-known paradigmatic models as the Lorenz's [Lorenz (1963)], Chua's [Matsumoto (1984); Chua *et al.* (1986)] Chen's [Hau *et al.* (2010); Celikovshy & Chen (2002)], Roessler's [Roessler (1976)] and other simple three-dimensional systems reported by Sprott [Sprott (2003)]. The Lorenz model is formally equivalent to the laser equations (the so-called class C-laser), as demonstrated by H. Haken [Haken (1975)] so the link with laser dynamics is direct. The intrinsic symmetries imply a chaotic trajectory visiting two saddle foci to be contrasted with the laser with feedback where the competing steady states are a saddle focus and a saddle node which contribute to the creation of the homoclinic connection. The comparison with the Chua's circuit which relies on a "locally active resistor" with static nonlinear characteristic is of another kind. This element is the Chua's diode and it can be implemented in different ways. The chaotic attractors from the Chua circuit perfectly reproduce Lorenz chaos even though the correspondence with laser equations is difficult to draw. The chaotic Chen system is similar but not equivalent to the Lorenz one as recently pointed out by Chen [Chen (2020)] introducing the concept of "generalized Lorenz systems family". In both cases the nonlinearities are two of the quadratic type in models which have seven terms on the right-hand side. The Roessler circuit is simpler when compared with Lorenz systems because it possesses only one quadratic term and seven terms on the right-hand side. It does not seem that both Chua and Roessler systems are related to laser dynamics. The above cited systems played a crucial role in chaos synchronization demonstrated by Pecora and Carroll [Pecora & Carroll (1990)] using Lorenz and Roessler circuits. The increased complexity in the laser with feedback is due to feedback process which implies the additional cubic term xz^2 to the two quadratic nonlinearities xy. This is the first time that the cubic term is treated in its simplest form approximating $\sin(z)^2$ with z^2. It is important to stress that adiabatic elimination of the fast variable corresponding to laser polarization in the Lorenz system does not lead to rate model of the laser equations in the so-called class B laser, the large class of the available lasers, including semiconductor lasers and solid state lasers. As far as laser semiconductor lasers are concerned, it is important to note that their dynamics is well described by the Lang and Kobayashi (LK) model [Lang & Kobayashi (1980)] accounting for the effects of delayed optical feedback acting on the timescale of the intrinsic semiconductor laser. The LK equations describe the complex dynamics of the complex electric field E and the inversion (number of electron-hole pairs) N inside the laser. The fast chaotic dynamics from these lasers has been largely used in applications for secure communication systems ([Fischer *et al.* (2000); Van Wiggeren & Roy (1998); Donati & Mirasso (2002); Ohtsubo & Davis (2005)]).

3.5 Conclusions

In conclusion, we retain that the laser with feedback, whose historical origins dates back in the same period of other chaotic oscillators will be appropriate for describing instances of local chaos, reached after subharmonic bifurcations, global bifurcations as homoclinic chaos and relaxation type oscillation behavior or regular spiking behavior when a control parameter is changed. Another valuable advantage over the nonlinear oscillators described above is related to the fact that elimination of the feedback variable leaves unchanged its potentialities, and chaos can be reached by modulation of the parameter k, provided the general condition on the timescales are fulfilled. In this framework, interesting perspectives are also foreseen for competition population dynamics ruled by Volterra–Lotka models [Volterra (1926, 1931); Lotka (1910, 1920)] and chaotic epidemiological models [Schwartz & Smith (1983)], in both cases the underlying nonlinearities are of the direct product of two variables.

3.6   Appendix

This appendix presents a method allowing to provide an upper bound for
the Hopf bifurcation parameter for any three-dimensional autonomous dynamical system for which the fixed point coordinates cannot be easily expressed analytically as it is the case for the dynamical system (3.2) for
which the coordinates of the fixed point I_2 are the roots of a cubic polynomial. Let us suppose that the three eigenvalues of the Jacobian matrix
J of this dynamical system evaluated at the fixed point (I_2 in our case)
are real and complex conjugate λ_1, $\lambda_{2,3} = \alpha \pm i\omega$. The Cayley–Hamilton
eigenpolynomial reads:

$$\lambda^3 - \sigma_1\lambda^2 + \sigma_2\lambda - \sigma_3 = 0 \tag{3.9}$$

where $\sigma_1 = Tr(J)$, $\sigma_2 = \sum_{i=1}^{3} M_{ii}(J)$ is the sum of all first-order diagonal
minors of J and $\sigma_3 = Det(J)$. Thus, we have:

$$\sigma_1 = Tr(J) = \lambda_1 + \lambda_2 + \lambda_3 = \lambda_1 + 2\alpha,$$

$$\sigma_2 = \sum_{i=1}^{3} M_{ii}(J) = \lambda_1\lambda_2 + \lambda_1\lambda_3 + \lambda_2\lambda_3 = 2\alpha\lambda_1 + \beta, \tag{3.10}$$

$$\sigma_3 = Det(J) = \lambda_1\lambda_2\lambda_3 = \lambda_1\beta,$$

where $\beta = \alpha^2 + \omega^2$. In order to analyze the stability of fixed points according to a control parameter value (B_0 here), i.e. the occurrence of Hopf
bifurcation, we propose to use the Routh–Hurwitz' theorem [Routh (1877);
Hurwitz (1893)] which states that if $D_1 = \sigma_2$ and $D_2 = \sigma_3 - \sigma_2\sigma_1$ are both
positive then *eigenpolynomial* equation (3.9) would have eigenvalues with
negative real parts. From Eq. (3.10) it can be stated that:

$$\alpha = \frac{\sigma_1\sigma_2 - \sigma_3}{\lambda_1^2 + \sigma_2} \tag{3.11}$$

Thus, $\alpha = 0$ provided that $\kappa = \sigma_1\sigma_2 - \sigma_3 = 0$. For dynamical system
(3.2), we obtain:

$$\kappa = (1+x)\left[p_0xy\epsilon_1 + (1+x+\epsilon_2)\epsilon_2\right] - 2B_1k_1xz\epsilon_1\epsilon_2^2 \tag{3.12}$$

then, by replacing x, y and z the coordinates (3.3) of the fixed point I_2, we
have:

$$B_0 = \frac{1+\epsilon_2}{2B_1k_1x^*\epsilon_1\epsilon_2} + \frac{p_0\epsilon_1 + \epsilon_2(2+\epsilon_2)}{2B_1k_1\epsilon_1\epsilon_2^2} + \frac{(1+2B_1^2k_1\epsilon_1\epsilon_2)x^*}{2B_1k_1\epsilon_1\epsilon_2} \tag{3.13}$$

Positivity of fixed points has led to $x^* \leqslant p_0 - 1$ which implies that $\max(x^*) = p_0 - 1$. Thus, by posing $x^* = p_0 - 1$ in (3.13) and while using the parameter sets of our experiment, i.e. $\epsilon_1 = 200$, $\epsilon_2 = 6$, $k_1 = 12$, $p_0 = 1.208$ and $B_1 = 0.555$, we find:

$$B_0^{Hopf} \leqslant 0.12057$$

The numerical computation of the Hopf bifurcation parameter value has been found to be equal to 0.12036 which is below and very near the upper bound analytically obtained.

3.7 References

Andronov A., Leontovich E., Gordon I. & Maier A. (1971). *Theory of Bifurcations of Dynamical Systems on a Plane*, Israel Program for Scientific Translations, Jerusalem.

Arecchi F.T., Meucci R., Puccioni G.P. & Tredicce J.R. (1982). Experimental evidence of subharmonic bifurcations, multistability, and turbulence in a Q-switched gas laser, *Physical Review Letters*, 49(17), pp. 1217-1220.

Arecchi F.T., Gadomski W. & Meucci R. (1986). Generation of chaotic dynamics by feedback on a laser, *Phys. Rev. A*, 34(2), pp. 1617-1620.

Arecchi F.T., Meucci R. & Gadomski W. (1987). Laser dynamics with competing instabilities, *Phys. Rev. Lett.*, 58(21), pp. 2205-2208.

Arecchi F.T., Fortuna L., Frasca M., Meucci R. & Sciuto G. (2005). A programmable electronic circuit for modelling CO_2 laser dynamics, *Chaos*, 15, 043104.

Celikovshy S. & Chen G. (2002). On a generalized Lorenz canonical form of chaotic systems, *Int. J. Bifurcation Chaos*, 12, pp. 1789-1812.

Chen G. (2020). Generalized Lorenz systems family, arXiv:2006.04066.

Chua L.O., Kumaro M. & Matsumoto T. (1986). The double scroll family, *IEEE Transactions on Circuits and Systems*, 33, pp. 1072-1118.

Ciofini M., Politi A. & Meucci R. (1993). Effective two-dimensional model for CO_2 lasers, *Phys. Rev. A*, 48, pp. 605-610.

Donati S. & Mirasso C.R. (2002). Feature section on optical chaos and applications to cryptography, *IEEE J. Quantum Electron.*, 38, pp. 1138-1205.

Eckmann J.P. & Ruelle D. (1985). Ergodic theory of chaos and strange attractors, *Rev. Mod. Phys.*, 57, pp. 617-656.

Fischer I., Liu Y. & Davis P. (2000). Synchronization of chaotic semiconductor laser dynamics on subnanosecond time scales and its potential for chaos communication, *Phys. Rev. A*, 62(1), 011801 (4 pages).

Freire J.G., Meucci R., Arecchi F.T. & Gallas J.A.C. (2015). Self-organization of pulsing and bursting in a CO_2 laser with opto-electronic feedback,

Chaos, 25(9), 097607.

Haken H. (1975). Analogy between higher instabilities in fluids and lasers, *Physics Letters A*, 53(1), pp. 77-78.

Hau Z., Kang N., Kong X., Chen G. & Yan G. (2010). On the equivalence of Lorenz system and Chen system, *Int. J. Bifurcation Chaos*, 20, pp. 557-560.

Hopf E. (1942). Abzweigung einer periodischen Lösung von einer stationären Lösung eines Differentialsystems, *Berichte der MathematischPhysikalischen Klasse der Sächsischen Akademie der Wissenschaften zu Leipzig*, Band XCIV, Sitzung vom 19. January 1942, pp. 3-22. See L.N. Howard and N. Kopell, A Translation of Hopf's Original Paper, pp. 163-193 and Editorial Comments, pp. 194-205 in J. Marsden and M. McCracken.

Hurwitz A. (1893). Über die Bedingungen, unter welchen eine Gleichung nur Wurzeln mit negativen reellen Theilen besitzt, *Math. Ann.*, 41, pp. 403-442.

Klein M. & Baier G. (1991). Hierarchies of dynamical systems, In *A Chaotic Hierarchy*, eds. G. Baier and M. Klein (World Scientific, Singapore.

Kuznetsov Yu. A. (2004). *Elements of Applied Bifurcation Theory* (Springer-Verlag, New York), Third edition.

Lang R. & Kobayashi K. (1980). External optical feedback effects on semiconductor injection laser properties, *IEEE Journal of Quantum Electronics*, 16(3), pp. 347-355.

Lorenz E.N. (1963). Deterministic non-periodic flows, *J. Atmos. Sci.*, 20, pp. 130-141.

Lotka A.J. (1910). Contribution to the theory of periodic reaction, *J. Phys. Chem.*, 14(3), pp. 271-274.

Lotka A.J. (1920). Analytical note on certain rhythmic relations in organic systems, *Proc. Natl. Acad. Sci. U.S.A.*, 6(7) pp. 410-415.

Marsden J. & McCracken M. (1976). *Hopf Bifurcation and its Applications* (Springer-Verlag, New York).

Matsumoto T. (1984). A chaotic attractor from Chua's circuit, *IEEE Transactions on Circuits and Systems*, 31, pp. 1055-1058.

Ohtsubo J. & Davis P. (2005). Chaotic optical communication, In *Unlocking Dynamical Diversity-optical Feedback Effects on Semiconductor Lasers*, Chap. 10, eds. D. Kane and K.A. Shore (Wiley, Chichester).

Pecora L.M. & Carroll T.L. (1990). Synchronization in chaotic systems, *Phys. Rev. Lett.*, 64, pp. 821-824.

Roessler, O.E. (1976). An equation for chaos, *Phys. Lett. A*, 57(5), pp. 397-398.

Routh, E.J. (1877). *A Treatise on the Stability of a Given State of Motion: Particularly Steady Motion* (Macmillan and Co.).

Sandri, M. (1996). Numerical calculation of Lyapunov exponents, *The Mathematica Journal*, 6(3), pp. 78-84.

Schwartz I.B. & Smith H. (1983). Infinite subharmonic bifurcation in an SEIR epidemic model, *Journal of Mathematical Biology*, 18(3), pp. 233-253.

Sprott, J.C. (2003). *Chaos and Time-Series Analysis* (Oxford University Press).

Van Wiggeren G.D. & Roy R. (1998). Communication with chaotic lasers, *Science*, 279, pp. 1198-1200.

Varone A., Politi A. & Ciofini M. (1995). CO_2 laser dynamics with feedback, *Phys. Rev. A*, 52, pp. 3176-3182.

Volterra, V. (1926). Variazioni e fluttuazioni del numero d'individui in specie animali conviventi, *Mem. Acad. Lincei Roma*, 2, pp. 31-113.

Volterra, V. (1931). Variations and fluctuations of the number of individuals in animal species living together, In *Animal Ecology*, ed. R.N. Chapman (McGraw-Hill, New York), pp. 409-448.

Weiss C.O. & Brock J. (1986). Evidence for Lorenz-type chaos in a laser, *Physical Review Letters*, 57(22), pp. 2804-2806.

Wolf A., Swift J.B., Swinney H.L. & Vastano J.A. (1985). Determining Lyapunov exponents from a time series, *Physica D*, 16, pp. 285-317.

Chapter 4

Slow Invariant Manifold of Laser with Feedback

Previous works have demonstrated experimentally and theoretically the existence of *slow fast* evolutions, i.e. slow chaotic spiking sequences in the dynamics of a semiconductor laser with ac-coupled optoelectronic feedback. In this chapter, by using the so-called *Flow Curvature Method*, we provide the *slow invariant manifold* analytical equation of such a laser model. This equation and its graphical representation in the phase space enable on the one hand to discriminate the slow evolution of the trajectory curves from the fast one and, on the other hand to improve our understanding of this *slow-fast* regime.

4.1 Introduction

More than ten years ago, Al-Naimee *et al.* [Al-Naimee *et al.* (2009)] studied "the occurrence of chaotic spiking in a semiconductor laser with ac-coupled nonlinear optoelectronic feedback. The solitary laser dynamics is ruled by two coupled variables (intensity and population inversion) evolving with two very different characteristic timescales. The introduction of a third degree of freedom (and a third timescale) describing the ac-feedback loop, leads to a three-dimensional slow–fast system displaying a transition from a stable steady state to periodic spiking sequences as the dc-pumping current is varied (...). The timescale of these dynamics, much slower with respect to typical semiconductor laser timescales (few *ns*), is fully determined by the highpass filter in the feedback loop." Then, they provided a minimal physical model qualitatively reproducing the experimental results. Since this model involves two time scales, it can be considered a *slow–fast dynamical system* or a *singularly perturbed system*.

At the end of the nineteenth century, Henri Poincaré originally developed, in his *New Methods of Celestial Mechanics* [Poincaré (1892)], *singular perturbation methods*. During the thirties and in the following decades, Andronov and Chaikin [Andronov (1937)], Levinson [Levinson (1949)] and Tikhonov [Tikhonov (1948)] generalized Poincaré's ideas and stated that *slow–fast dynamical systems*, also called *singularly perturbed systems*, possess *invariant manifolds* on which trajectories evolve slowly, and toward which nearby orbits contract exponentially in time (either forward or backward) in the normal directions. Then, from the beginning of the sixties, the seminal works of Wasow [Wasow (1965)], Cole [Cole (1968)], O'Malley [O'Malley (1974, 1991)] and Fenichel [Fenichel (1971, 1974, 1977, 1979)], to name but a few, gave rise to the so-called *Geometric Singular Perturbation Theory*. Fenichel [Fenichel (1971, 1974, 1977, 1979)] established the *local invariance* of *slow invariant manifolds* that possess both expanding and contracting directions and which were labeled *slow invariant manifolds* while using his theory for the *persistence of normally hyperbolic invariant manifolds*. Let us note that the theory of *invariant manifolds* for an ordinary differential equation was independently developed by Hirsch, *et al.* [Hirsch (1977)]. Since the beginning of the eighties, two kinds of approaches have been developed: *singular perturbation-based methods* and *curvature-based methods*. The former include the *Geometric Singular Perturbation Theory* (GSPT), the *Successive Approximations Method* (SAM) [Rossetto (1986, 1987)] and the *Zero-Derivative Principle* (ZDP) [Gear *et al.* (2005); Zagaris *et al.* (2009)], and the latter, the *Intrinsic Low-Dimensional Manifold* (ILDM) [Maas & Pope (1992)], the *Inflection Line Method* (ILM) [Brøns & Bar-Eli (1994)] and the *Tangent Linear System Approximation* (TLSA) [Rossetto *et al.* (1998)]. In 2005, a new approach of n-dimensional *singularly perturbed dynamical systems* or *slow–fast dynamical systems* based on the location of the points where the *curvature* of *trajectory curves* vanishes, called the *Flow Curvature Method*, was developed by Ginoux *et al.* [Ginoux (2006a,b, 2008)] and then by Ginoux [Ginoux (2009, 2011, 2014)]. In a recent publication, Ginoux [Ginoux (2021)] proved, on the one hand, the identity between all the methods belonging to the same category (i.e. belonging to *singular perturbation-based methods* or to *curvature-based methods*) and, on the other hand, that between both categories. Moreover, he also established, on the one hand, that his *Flow Curvature Method* encompasses the three other methods (IDLM, TLSA, and ILM) and, on the other hand, the identity between his *Flow Curvature Method* and *Geometric Singular Perturbation Method*.

Thus, the aim of this work was to provide the *slow invariant manifold* analytical equation of the semiconductor laser model with ac-coupled optoelectronic feedback introduced by Al-Naimee *et al.* [Al-Naimee *et al.* (2009)]. Let us observe that, since this three-dimensional model has two small multiplicative parameters on the right-hand side of its velocity vector field, it has two *fast* variables and one *slow*. Thus, as highlighted by Ginoux and Rossetto [Ginoux (2006a)], in this specific case, one of the hypotheses of Tihonov's theorem [Tikhonov (1948)] is not checked since the *fast* dynamics of the *singular approximation*, i.e. the zero-order approximation of the *slow invariant manifold*, has a periodic solution. As a consequence, *Geometric Singular Perturbation Theory* fails to provide its analytical equation. To overcome such difficulty, we used, in this work, the so-called *Flow Curvature Method*. This chapter is organized as follows: in Section 4.2, the laser model and its parameters are presented. In Section 4.3, the main features of the *Flow Curvature Method* are recalled and the *slow invariant manifold* of the laser model is provided as well as its graphical representation in the phase space. In the last section, the results are discussed, and perspectives on this work are given.

4.2 Slow–fast dynamical system

Following the work of Al-Naimee *et al.* [Al-Naimee *et al.* (2009)], we will use the dynamical system:

$$\frac{dx}{dt} = x\,(y-1)\,,$$

$$\frac{dy}{dt} = \nu\,(\delta_0 - y + f\,(x,z) - xy)\,, \qquad (4.1)$$

$$\frac{dz}{dt} = -\varepsilon\,(x+z)\,,$$

where

$$f\,(x,z) = \alpha\frac{x+z}{1+s\,(x+z)} \qquad (4.2)$$

and the parameters $s = 11$, $\alpha = 1$, $\nu = 10^{-3}$, and $\varepsilon = 2 \times 10^{-5}$ as well as the *bifurcation parameter* $\delta_0 = 1.017$ are exactly the same as in [Al-Naimee *et al.* (2009)]. Due to the presence of the two small parameters ν and ε, the dynamical system (4.1) is considered *slow–fast*. However, let us observe that $\varepsilon = \kappa\nu$ where $\kappa = 2 \times 10^{-2}$. Thus, by posing $\tau = \nu t$, system (4.1) reads:

$$\frac{dx}{d\tau} = x' = \frac{x}{\nu}(y-1),$$

$$\frac{dy}{d\tau} = y' = \delta_0 - y + f(x,z) - xy, \qquad (4.3)$$

$$\frac{dz}{d\tau} = z' = -\kappa(x+z),$$

Thus, $x' = O(\nu^{-1}) \gg 1$, since $\nu \ll 1$, $y' = O(1)$, and $z' = O(\kappa) \ll 1$, since $\kappa \ll 1$. It follows that x is a (very) *fast* variable and y is a *fast* variable, while z is a *slow* variable.

4.3 Stability analysis

4.3.1 *Fixed points, Jacobian matrix and eigenvalues*

Fixed points are determined while using the classical nullclines method. System (4.1) has two fixed points: $I_1(0, \delta, 0)$ and $I_2(\delta - 1, 1, 1 - \delta)$.

The Jacobian matrix of system (4.1) reads:

$$J = \begin{pmatrix} y-1 & x & 0 \\ \nu\left(\dfrac{\partial f}{\partial x} - y\right) & -\nu(1+x) & \nu\dfrac{\partial f}{\partial z} \\ -\varepsilon & 0 & -\epsilon \end{pmatrix} \qquad (4.4)$$

Let us observe that for both the fixed points I_1 and I_2:

$$\frac{\partial f}{\partial x} = \frac{\alpha}{[1 + s(x+z)]^2} = \frac{\partial f}{\partial z} = \alpha = 1, \qquad (4.5)$$

since nullcline $\dot{z} = -\varepsilon(x+z) = 0$ and the parameter $\alpha = 1$.

By replacing the coordinate of the fixed point I_1 in the Jacobian matrix (4.4), the Cayley–Hamilton eigenpolynomial can be easily factorized and reads:

$$[\lambda - (\delta - 1)](\lambda + \nu)(\lambda + \varepsilon) = 0 \qquad (4.6)$$

Thus, there are three real eigenvalues:

$$\lambda_1 = \delta - 1 \quad ; \quad \lambda_2 = -\nu \quad ; \quad \lambda_3 = -\varepsilon. \qquad (4.7)$$

Thus, the fixed point I_1 is a *saddle-node* provided that $\delta > 1$.

Then, upon replacing the coordinate of the fixed point I_2 in the Jacobian matrix (4.4), the Cayley–Hamilton eigenpolynomial reads:

$$\lambda^3 - (\nu\delta + \varepsilon)\,\lambda^2 + \nu\delta\varepsilon\lambda - (1 - \delta)\,\nu\varepsilon = 0 \tag{4.8}$$

Numerically solving this third-degree eigenpolynomial (4.8) leads to:

$$\lambda_1 = -0.00124128 \quad ; \quad \lambda_{2,3} = 0.000102141 \pm 0.000513301i. \tag{4.9}$$

Thus, the fixed point I_2 is a *saddle-focus*.

By using *perturbation methods* [Bender & Orszag (1999)], the real root λ_1 of the eigenpolynomial (4.8) may be approximated by:

$$\lambda_1 = -\nu\delta + O\left(\varepsilon\right), \tag{4.10}$$

where $O\left(\varepsilon\right) = -k\varepsilon$ with $k \gg \varepsilon$. Moreover, the trace of the Jacobian matrix (4.4), evaluated at the fixed point I_2, provides:

$$Tr\left(J\right) = \lambda_1 + \lambda_2 + \lambda_3 = -\nu\delta - \varepsilon. \tag{4.11}$$

Since $\lambda_{2,3} = \sigma \pm i\omega$ is a complex conjugate, this trace reads:

$$Tr\left(J\right) = \lambda_1 + 2\sigma = -\nu\delta - \varepsilon. \tag{4.12}$$

Thus, by replacing, in the previous equation (4.12), λ_1 with the expression (4.10), we obtain:

$$2\sigma = \left(k - 1\right)\varepsilon > 0. \tag{4.13}$$

It follows that the real part of the eigenvalues $\lambda_{2,3}$ is necessarily positive and, thus, the fixed point I_2 is a *saddle-focus*.

4.3.2 *Bifurcation diagram*

Following the work of Al-Naimee *et al.* [Al-Naimee *et al.* (2009)], we used δ as a bifurcation parameter and computed the bifurcation diagram, which is presented in Figure 4.1. Such a diagram, which is exactly the same as that produced in [Al-Naimee *et al.* (2009)], provides information that can be used to have a better understanding of the phase space orbits plotted in Figure 4.2.

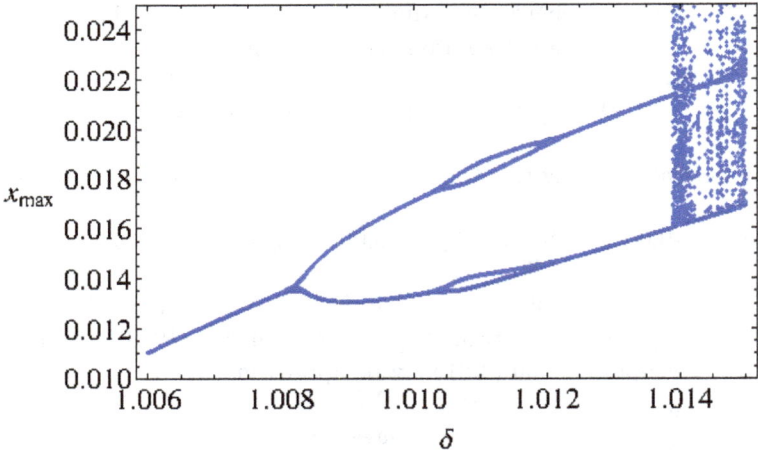

Figure 4.1. Bifurcation diagram x_{max} as a function of δ.

From Figure 4.1, we observe that, for $\delta \in [1.006, 1.0082]$, the attractor is a *limit cycle* of period one (see Figure 4.2a). For $\delta \in [1.0082, 1.0105]$ and $\delta \in [1.0117, 1.0137]$, the period of the *limit cycle* becomes of period two (see Figure 4.2b). For $\delta \in [1.0105, 1.0117]$, the *limit cycle* is of period four (see Figure 4.2c). When $\delta > 1.037$, the attractor becomes chaotic (see Figure 4.2d). To confirm these results, Lyapunov characteristic exponents were computed in each case.

4.3.3 *Numerical computation of the Lyapunov characteristic exponents*

The numerical computation of the Lyapunov Characteristic Exponents (LCEs) of the system (4.1) was performed in each case with the algorithm developed by Sandri [Sandri (1996)] for Mathematica® and the Lyapunov Exponents Toolbox (LET) developed by Siu for MatLab® and involving the two algorithms proposed by Wolf *et al.* [Wolf *et al.* (1985)] and Eckmann and Ruelle [Eckmann & Ruelle (1985)] (see https://fr.mathworks.com/matlabcentral/fileexchange/233-let). The results obtained by both algorithms are consistent. The LCE values were computed within each considered interval $\delta \in [1.006, 1.02]$. As an example, for $\delta = 1.007, 1.010, 1.011$ and 1.017, both algorithms provided, respectively, the following LCEs: $(0, -0.15, -0.82)$, $(0, -0.17, -0.81)$, $(0, -0.16, -0.80)$

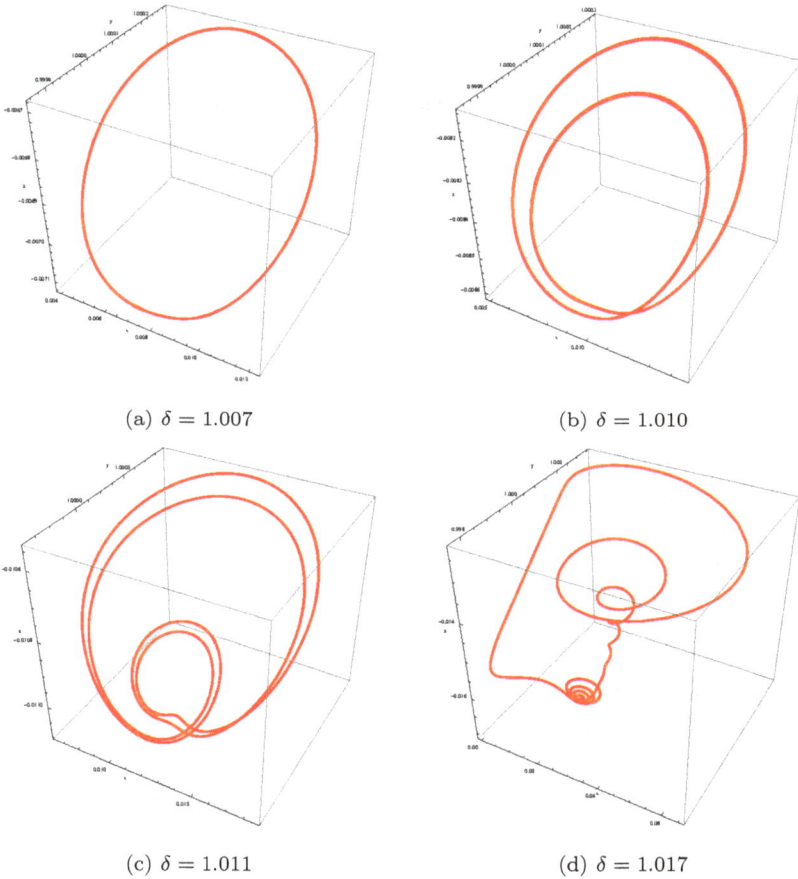

(a) $\delta = 1.007$

(b) $\delta = 1.010$

(c) $\delta = 1.011$

(d) $\delta = 1.017$

Figure 4.2.   Phase portraits of system (4.1) in the phase space for various values of δ.

and $(0.025, 0, -1.06)$. Then, the classification of (autonomous) continuous-time attractors of the dynamical system (4.6) on the basis of their Lyapunov spectrum, together with their Hausdorff dimension, is presented in Table 4.1 according to the work of Klein and Baier [Klein & Baier (1991)].

4.4   Slow invariant manifold

In recent publications, a new approach to n-dimensional *singularly perturbed systems* of ordinary differential equations, called the *Flow Curvature*

Table 4.1. Lyapunov characteristic exponents of dynamical system (4.6) for various values of δ.

δ	LCE spectrum	Dynamics of the attractor	Hausdorff dimension
$\delta \in [1.0060, 1.0082]$	$(0, -, -)$	Limit Cycle of period 1	$D = 1$
$\delta \in [1.0082, 1.0105]$	$(0, -, -)$	Limit Cycle of period 2	$D = 1$
$\delta \in [1.0105, 1.0117]$	$(0, -, -)$	Limit Cycle of period 4	$D = 1$
$\delta \in [1.0117, 1.0137]$	$(0, -, -)$	Limit Cycle of period 2	$D = 1$
$\delta \in [1.0137, 1.02]$	$(+, 0, -)$	2-Chaos	$D = 2.02$

Method, has been developed by Ginoux *et al.* [Ginoux (2006a,b, 2008, 2009, 2011, 2013, 2014, 2015, 2016, 2019)]. It considers the *trajectory curves* integral of such systems as curves in Euclidean n-space. Based on the use of local metric properties of *curvatures* inherent to *Differential Geometry*, this method, which does not require the use of asymptotic expansions, states that the location of the points where the local *curvature* of the *trajectory curves* of such systems is null defines an $(n-1)$-dimensional *manifold* associated with this system and called the *flow curvature manifold*. The invariance of this manifold is then stated according to a theorem introduced by Gaston Darboux [Darboux (1878)] in 1878. Moreover, as stated in Ginoux [Ginoux (2009)], if the *slow–fast dynamical system* has a *symmetry* such as $(-x, -y, z)$, its *flow curvature manifold* has the same, i.e. $\phi(-x, -y, z) = \phi(x, y, z)$. Thus, as previously stated (see Section 4.2), the system (4.1) is a three-dimensional *singularly perturbed dynamical system* with *two* fast variables. However, in such a specific case, one of the hypotheses of Tikhonov's theorem [Tikhonov (1948)] is not checked since the *fast dynamics* of the *singular approximation* have a periodic solution. Nevertheless, while using the *Flow Curvature Method*, an approximation up to order three in $\nu\varepsilon^2$ and $\nu^2\varepsilon$ of the *slow invariant manifold* equation of the system (4.1) can been computed for various values of the *bifurcation parameter* δ.

According to the *Flow Curvature Method*, the following proposition holds for any n-dimensional *singularly perturbed dynamical system* comprising small multiplicative parameters in its velocity vector field:

Proposition 4.1. *The location of the points where the $(n-1)^{th}$ curvature of the flow, i.e. the curvature of the trajectory curve $\vec{X}$ integral of any n-dimensional singularly perturbed dynamical systems vanishes, providing* *a p-order approximation of its slow manifold, the equation of which reads*

$$\phi(\vec{X}) = \dot{\vec{X}} \cdot (\ddot{\vec{X}} \wedge \dddot{\vec{X}} \wedge \ldots \wedge \overset{(n)}{\vec{X}}) = \det(\dot{\vec{X}}, \ddot{\vec{X}}, \dddot{\vec{X}}, \ldots, \overset{(n)}{\vec{X}}) = 0, \qquad (4.14)$$

where $\overset{(n)}{\vec{X}}$ represents the time derivatives up to order n of the velocity vector field.

For a proof of this proposition and that of the invariance of the *slow manifold* (4.14), see Ginoux [Ginoux (2006a,b, 2008, 2009, 2011, 2013, 2014, 2021)]. Let us observe that the *p*-order approximation depends on the number of small multiplicative parameters in the velocity vector field. In the particular case of system (4.1), it can be easily stated that, since the plane $x = 0$ is *invariant* with respect to the flow and due to the presence of two small multiplicative parameters, i.e. ν and ε, the *slow invariant manifold* equation (4.14) can be simply and directly expressed by:

$$\phi(\vec{X}) = (\dot{\vec{X}} \wedge \ddot{\vec{X}}) \cdot \vec{i} = 0, \qquad (4.15)$$

where $\vec{i}$ is the unit vector along the x-axis. Thus, by using the *Flow Curvature Method* and Eq. (4.15), the *slow invariant manifold* equation of system (4.1) reads:

$$\begin{aligned}
\phi(x, y, z, \nu, \varepsilon) &= x(1 - y)(\delta + \delta s^2 x^2 + s^2 x^2 yz - s^2 x^2 y + 2s^2 xyz^2 - 2s^2 xyz + 2\delta s^2 xz \\
&+ s^2 yz^3 - s^2 yz^2 + \delta s^2 z^2 + sx^2 + 2\delta sx + 2sxyz - 2sxy + 2sxz + 2syz^2 - 2syz \\
&+ sz^2 + 2\delta sz + yz - y) - \epsilon(x + z)(-\delta + s^2 x^3 y - \delta s^2 x^2 + 2s^2 x^2 yz + s^2 x^2 y \\
&+ s^2 xyz^2 + 2s^2 xyz - 2\delta s^2 xz + s^2 yz^2 - \delta s^2 z^2 + 2sx^2 y - sx^2 - 2\delta sx + 2sxyz \\
&+ 2sxy - 2sxz + 2syz - sz^2 - 2\delta sz + xy + y) + \nu(x + 1)(x + z)(sx + sz + 1) \\
&\times (-\delta + sx^2 y - \delta sx + sxyz + sxy + syz - \delta sz + xy - x + y - z). \qquad (4.16)
\end{aligned}$$

The *slow invariant manifold* equations (4.16) of the system (4.1) have been plotted in the figures below (see Figure 4.3) for various values of the *bifurcation parameter* $\delta = 1.007, 1.010, 1.011$ and 1.017.

We observe that, for $\delta = 1.007$, the *trajectory curve* integral of system (4.1) is a periodic *stable limit cycle* that lies partly on the left side and right side of the *slow invariant manifold*; see Figure 4.3a. For a *bifurcation parameter* δ equal to 1.010 and 1.011 (see Figures 4.3b and 4.3c), the same evolution appears. Let us observe that the kind of funnel in Figures 4.3a and 4.3c corresponds to the attractive eigendirection of the fixed point I_2.

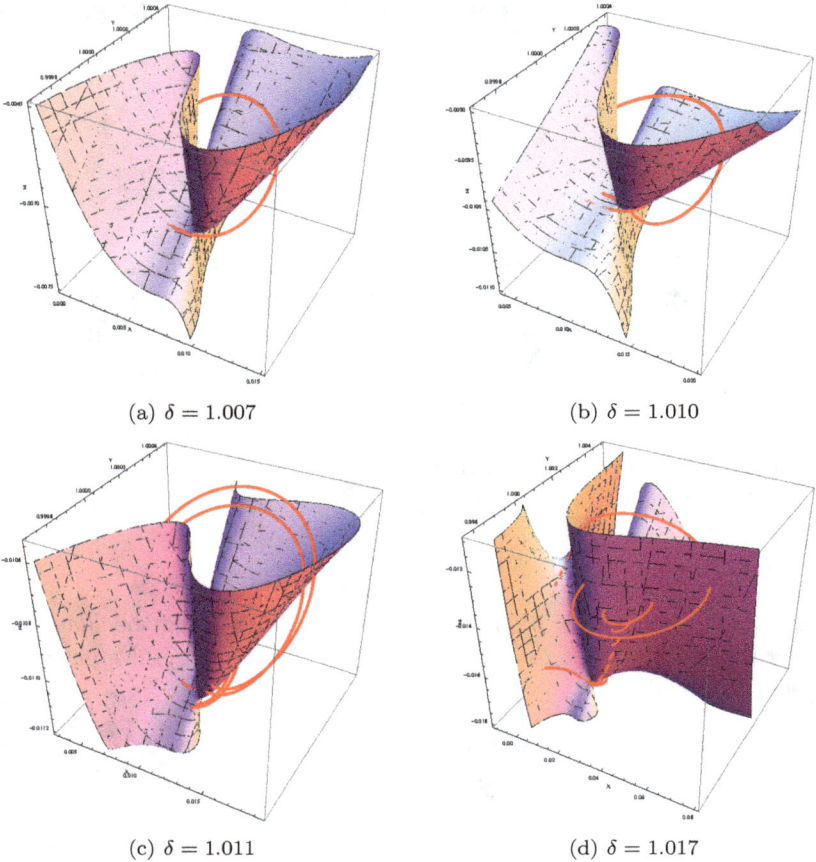

(a) $\delta = 1.007$

(b) $\delta = 1.010$

(c) $\delta = 1.011$

(d) $\delta = 1.017$

Figure 4.3.   Slow invariant manifolds of system (4.1) in the phase space for various values of δ.

When $\delta = 1.017$, the attractor becomes chaotic and the *trajectory curve* evolves slowly from the bottom to the top on nearly all the left part of the *slow invariant manifold*. Then, it jumps on the upper right part of *slow invariant manifold* and starts spiraling around the attractive eigendirection corresponding to the negative real eigenvalues $\lambda_1 \approx -\nu\delta$ of fixed point I_2; see Eqs. (4.9) and (4.10). When the *trajectory curve* reaches the lower right part of the *slow invariant manifold*, it jumps to its left part. Let us observe that, during its descent, it lies in the vicinity of the *slow invariant manifold*; see Figures 4.3d and 4.4.

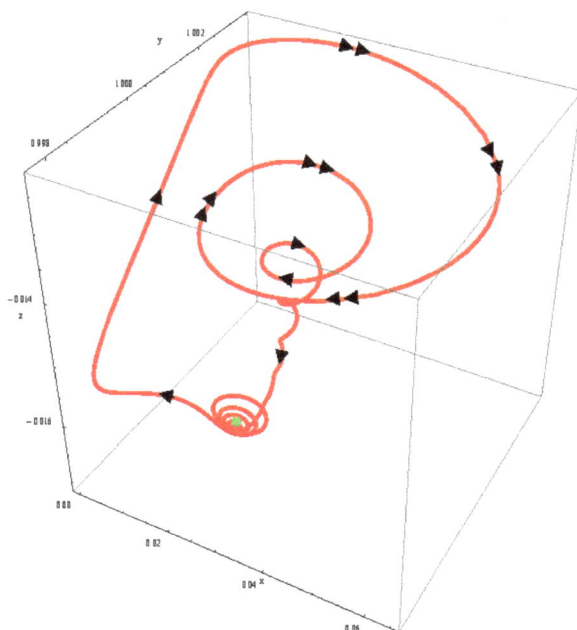

Figure 4.4. Evolution of the *trajectory curve* integral of system (4.1) for $\delta = 1.017$.

4.5 Discussion

In this work, by using the so-called *Flow Curvature Method*, we provide the *slow invariant manifold* analytical equation of a laser model. This equation and its graphical representation in the phase space enable, on the one hand, discriminating the *slow* evolution of the trajectory curves from the *fast* one and, on the other hand, improving our understanding of this *slow–fast* regime. Thus, the repelling and attracting branches of the *slow invariant manifold* have been specified. We also highlighted that the deformation of the surface representing, in the phase space (x, y, z), the slow invariant analytical manifold according to the bifurcation parameter δ is rather weak (see Figure 4.3). This confirms an assumption made by Al-Naimee *et al.* [Al-Naimee *et al.* (2009)] according to which:

> "Therefore, it is expected that they should not imply strong modifications of the slow-manifold shape which, as discussed above, is responsible for the observed dynamics."

Finally, our mathematical analysis also enabled confirming another assumption made by Al-Naimee *et al.* [Al-Naimee *et al.* (2009)], according to which "since (x_1, y_1, z_1) is located precisely on the slow manifold, the exact homoclinic connection does not occur."

4.6 References

Al-Naimee K., Marino F., Ciszak M., Meucci R. & Arecchi F.T. (2009). Chaotic spiking and incomplete homoclinic scenarios in semiconductor lasers with optoelectronic feedback, *New J. Phys.*, 11, 073022.

Poincaré H. (1892, 1893, 1899). *Les méthodes Nouvelles de la Mécanique Céleste* (Gauthier-Villars, Paris, France), Volumes I, II & III.

Andronov A.A. & Chaikin S.E. (1937). *Theory of Oscillators*, Moscow, I., English Translation; (Princeton University Press, Princeton, NJ, USA), 1949.

Levinson N. (1949). A second-order differential equation with singular solutions, *Ann. Math.*, 50, pp. 127-153.

Tikhonov A.N. (1948). On the dependence of solutions of differential equations on a small parameter, *Mat. Sb. N.S.*, 31, pp. 575-586.

Wasow W.R. (1965). *Asymptotic Expansions for Ordinary Differential Equations* (Wiley-Interscience, New York, NY, USA).

Cole J.D. (1968). *Perturbation Methods in Applied Mathematics* (Blaisdell, Waltham, MA, USA).

O'Malley R.E. (1974). *Introduction to Singular Perturbations* (Academic Press, New York, NY, USA).

O'Malley R.E. (1991). *Singular Perturbations Methods for Ordinary Differential Equations* (Springer, New York, NY, USA).

Fenichel N. (1971). Persistence and smoothness of invariant manifolds for flows, *Ind. Univ. Math. J.*, 21, pp. 193-225.

Fenichel N. (1974). Asymptotic stability with rate conditions, *Ind. Univ. Math. J.*, 23, pp. 1109-1137.

Fenichel N. (1977). Asymptotic stability with rate conditions II, *Ind. Univ. Math. J.*, 26, pp. 81-93.

Fenichel N. (1979). Geometric singular perturbation theory for ordinary differential equations, *J. Diff. Eq.*, 31, pp. 53-98.

Hirsch M.W., Pugh C.C. & Shub M. (1977). *Invariant Manifolds* (Springer, New York, NY, USA).

Rossetto B. (1986). Trajectoires lentes des systèmes dynamiques lents-rapides, In *Analysis and Optimization of System* (Springer, Berlin/ Heidelberg, Germany), pp. 680-695.

Rossetto B. (1987). Singular approximation of chaotic slow-fast dynamical systems, In *The Physics of Phase Space Nonlinear Dynamics and*

Chaos Geometric Quantization, and Wigner Function (Springer, Berlin/ Heidelberg, Germany), pp. 12-14.

Gear C.W., Kaper T.J., Kevrekidis I.G. & Zagaris A. (2005). Projecting to a slow manifold: Singularly perturbed systems and legacy codes, *SIAM J. Appl. Dyn. Syst. Math.*, 4, pp. 711-732.

Zagaris A., Gear C.W., Kaper T.J. & Kevrekidis Y.G. (2009). Analysis of the accuracy and convergence of equation-free projection to a slow manifold, *ESAIM Math. Model. Num.*, 43, pp. 757-784.

Maas U. & Pope S.B. (1992). Simplifying chemical kinetics: Intrinsic low-dimensional manifolds in composition space, *Combust. Flame*, 88, pp. 239-264.

Brøns M. & Bar-Eli K. (1994). Asymptotic analysis of canards in the EOE equations and the role of the inflection line, In *Proceedings of the Royal Society of London, Series A: Mathematical, Physical and Engineering Sciences*, Vol. 445, pp. 305-322.

Rossetto B., Lenzini T., Ramdani S. & Suchey G. (1998). Slow fast autonomous dynamical systems, *Int. J. Bifurcation Chaos*, 8, pp. 2135-2145.

Ginoux J.M. & Rossetto B. (2006). Slow manifold of a neuronal bursting model, In *Emergent Properties in Natural and Articial Dynamical Systems*, eds. M.A. Aziz-Alaoui and C. Bertelle (Springer, Berlin/ Heidelberg, Germany), pp. 119-128.

Ginoux J.M. & Rossetto B. (2006). Differential geometry and mechanics applications to chaotic dynamical systems, *Int. J. Bifurcation Chaos*, 4, pp. 887-910.

Ginoux J.M., Rossetto B. & Chua L.O. (2008). Slow invariant manifolds as curvature of the flow of dynamical systems, *Int. J. Bifurcation Chaos*, 11, pp. 3409-3430.

Ginoux J.M. (2009). *Differential Geometry Applied to Dynamical Systems* World Scientific Series on Nonlinear Science, Series A, Vol. 66 (World Scientific, Singapore).

Ginoux J.M. & Llibre J. (2011). The flow curvature method applied to canard explosion, *J. Phys. A Math. Theor.*, 44, 465203.

Ginoux J.M. (2014). The slow invariant manifold of the Lorenz-Krishnamurthy model, *Qual. Theory Dyn. Syst.*, 13, pp. 19-37.

Ginoux J.M. (2021). Slow invariant manifolds of slow-fast dynamical systems, *Int. J. Bifurcation Chaos*, 31, 2150112-1-17.

Bender C.M. & Orszag S.A. (1999). *Advanced Mathematical Methods for Scientists and Engineers* (Springer, New York, NY, USA).

Sandri M. (1996). Numerical calculation of Lyapunov exponents, *Math. J.*, 6, pp. 78-84.

Wolf A., Swift J.B., Swinney H.L. & Vastano J.A. (1985). Determining Lyapunov exponents from a time series, *Physica D*, 16, pp. 285-317.

Eckmann J.P. & Ruelle D. (1985). Ergodic theory of chaos and strange attractors, *Rev. Mod. Phys.*, 57, pp. 617-656.

Klein M. & Baier G. (1991). Hierarchies of dynamical systems, In *A Chaotic Hierarchy*, eds. G. Baier and M. Klein (World Scientific, Singapore).

Ginoux J.M., Llibre J. & Chua L.O. (2013). Canards from Chua's circuit, *Int. J. Bifurcation Chaos*, 23, 1330010.

Ginoux J.M. & Llibre J. (2015). Canards existence in FitzHugh–Nagumo and Hodgkin–Huxley neuronal models, *Math. Probl. Eng.*, 2015, 342010.

Ginoux J.M. & Llibre J. (2016). Canards existence in memristor's circuits, *Qual. Theory Dyn. Syst.*, 15, pp. 383-431.

Ginoux J.M., Llibre J. & Tchizawa K. (2019). Canards existence in the Hindmarsh–Rose model, *Math. Model. Nat. Phenom.*, 14, pp. 1-21.

Darboux G. (1878). Mémoire sur les équations différentielles algébriques du premier ordre et du premier degré, *Bull. Sci. Math.*, Sér. 2, p. 60-96, p. 123-143 & p. 151-200.

Chapter 5

Phase Control in Nonlinear Systems

5.1 Introduction

Since the pioneering work on controlling chaos due to Ott, Grebogi and Yorke (OGY) [Ott *et al.* (1990)], different control schemes have been proposed that allow to obtain a desired response from a dynamical system by applying some small but accurately chosen perturbations [Shinbrot *et al.* (1993)]. In this context, some techniques that allow avoiding escapes in open dynamical systems presenting transient chaos have been proposed, with applications to many different situations in physics and engineering (see Ref. [Aguirre *et al.* (2004)] and references therein).

The methods stated to control chaos can be classified in *feedback* and *nonfeedback methods* [Boccaletti *et al.* (2000)], depending on how they interact with the system. Feedback methods of chaos control, as the celebrated OGY [Ott *et al.* (1990)], stabilize one of the unstable orbits that lie in the chaotic attractor by using small state-dependent perturbations into the system. However, in experimental implementations, the fast response that these methods require cannot usually be provided. For these situations, nonfeedback methods are more useful. Nonfeedback methods have been mainly used to suppress chaos in periodically driven dynamical systems.

$$\dot{\mathbf{x}} = \mathbf{f}(\mathbf{x}, \lambda) + \mathbf{F} \cos \omega t \tag{5.1}$$

where $\mathbf{x}, \mathbf{f}$ and $\mathbf{F}$ are vectors of the m-dimensional phase space, and λ is a parameter of the system. The main idea of these nonfeedback methods is to apply a harmonic perturbation either to some of the parameters of the system

$$\dot{\mathbf{x}} = \mathbf{f}(\mathbf{x}, \lambda(1 + \epsilon \cos(r\omega t + \phi))) + \mathbf{F} \cos \omega t \tag{5.2}$$

or as an additional forcing,

$$\dot{\mathbf{x}} = \mathbf{f}(\mathbf{x}, \lambda) + \mathbf{F}\cos\omega t + \epsilon\mathbf{u}\cos(r\omega t + \phi) \qquad (5.3)$$

where $\mathbf{u}$ is a conveniently chosen unitary vector.

The effectiveness of this type of methods has been tested experimentally in different works [Lima *et al.* (1990); Meucci *et al.* (1994)]. In the first where these nonfeedback methods were explored, the numerical and experimental explorations were essentially focused on the role played by the perturbation amplitude ϵ and the resonance condition r, but the role of the phase difference ϕ was hardly explored. However, in [Meucci *et al.* (1994)], it was observed that the phase difference ϕ between the periodic forcing and the perturbation had certain influence on the dynamical behavior of the system. Furthermore, in [Qu *et al.* (1995)], the authors have shown that ϕ plays a crucial role on the global dynamics of the system. Thus, it was clear that the role of the phase difference is important in the global dynamics of the system. The type of control based on varying the phase difference ϕ in search of a desired dynamical behavior is known as the *phase control* technique.

The aim of this chapter is to show that the phase control is very versatile and that it can be applied in many different contexts. As we said, this technique can be used to control chaos, but also to control other paradigmatic dynamical behaviors present in nonlinear dynamical systems. Thus, we are going to show in this chapter that phase control can be used to suppress chaos [Zambrano *et al.* (2006)], but also to control the phenomenon of crisis-induced intermittency [Zambrano *et al.* (2006)] and to avoid escapes in an open dynamical system that presents transient chaotic behavior [Seoane *et al.* (2008)]).

This chapter is organized as follows. In Section 5.2 we present an application of the phase control method to the paradigmatic model for both the CO_2 laser and how this technique can be used to control crisis-induced intermittency. Some conclusions and a discussion of the main results of this chapter are presented in Section 5.3.

5.2 Phase control of intermittency in dynamical systems

5.2.1 *Crisis-induced intermittency and its control*

For some chaotic systems it can be observed that, by modifying a control parameter, the chaotic attractor can touch an unstable periodic orbit inside its basin of attraction. This phenomenon, known as *interior crisis* [Ott *et al.* (1982)], induces a sudden expansion of the attractor, after which most trajectories alternate over periods of time in the region where the pre-crisis attractor lay with excursions out of it.

This type of intermittency, called *crisis-induced intermittency* is a widespread phenomenon [Ott *et al.* (1987)], so it is common to find situations where a control of this type of behavior becomes desirable. Fifteen years ago, [Meucci *et al.* (2005)] a feedback method to enhance or tame the intermittency has been devised. The strategy is to force the system with a feedback in which the "typical" frequency of the excursions, that is the frequency of the periodic orbit involved in the interior crisis, is either filtered or enhanced. This method has been shown to be effective in a periodically driven chaotic CO_2 laser, as the one described in [Meucci *et al.* (2004)].

However, as we said in the previous section, feedback control methods might present some difficulties for their implementation. Thus, in some contexts nonfeedback methods might be more useful. In some of our previous works, we have shown that phase control of chaos [Qu *et al.* (1995); Yang *et al.* (1996)] is a powerful tool to control the dynamics of a periodically driven chaotic system. In this section, we are going to show that the intermittent behavior of a dynamical system close to an interior crisis can be controlled by using the phase control scheme. We give experimental and numerical evidence of the validity of the method for the periodically driven CO_2 laser close to an interior crisis, and in order to have a deeper insight on the role of ϕ we also present an analysis of phase control of the quadratic map close to a crisis.

5.2.2 *Experimental setup and implementation of the phase control scheme*

We first address the experimental implementation of the phase control scheme on a CO_2 laser, to control its intermittent behavior. The experimental setup consists of a single-mode CO_2 laser, as shown in Figure 5.1. The laser cavity is defined by a totally reflecting grating and a partially reflecting mirror (G and M), and the gain medium is pumped by a constant electric discharge current. An electro-optic modulator (EOM) is inserted in the laser cavity in order to control the cavity losses by an external forcing, obtained from a sinusoidal generator (MD), that can be represented as

$$F(t) = \beta \sin(2\pi f_0 t) + b_0, \qquad (5.4)$$

where β is the amplitude of the external forcing, b_0 is a bias voltage and $f_0 = 100$ kHz is about twice the relaxation frequency of the laser.

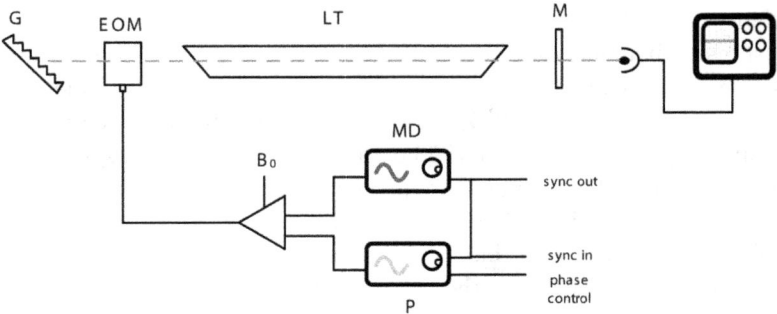

Figure 5.1.   Experimental setup for a single-mode CO_2 laser with modulated losses. EOM: intra-cavity electro-optic modulator, G: total reflecting grating, M: partial reflecting mirror, D: fast infrared detector, P: sinusoidal generator, MD: digital oscilloscope.

The CO_2 modulated laser is accurately described by the following model of five differential equations [Marino & Miguez (2006)]:

$$
\begin{aligned}
\dot{x}_1 &= kx_1(x_2 - 1 - \alpha \sin^2(F(t))) \\
\dot{x}_2 &= -\gamma_1 x_2 - 2kx_1 x_2 + gx_3 + x_4 + p \\
\dot{x}_3 &= -\gamma_1 x_3 + gx_2 + x_5 + p \qquad (5.5) \\
\dot{x}_4 &= -\gamma_2 x_4 + zx_2 + gx_5 + zp \\
\dot{x}_5 &= -\gamma_2 x_5 + zx_3 + gx_4 + zp.
\end{aligned}
$$

In the above equations, x_1 represents the laser output intensity, x_2 is the population inversion between the two resonant levels, and x_3, x_4 and x_5 account for molecular exchanges between the two levels resonant with the radiation field and the other rotational levels of the same vibrational band. The parameters of the model are the following: k is the unperturbed cavity loss parameter, g is a coupling constant, γ_1 and γ_2 are population relaxation rates, z accounts for an effective number of rotational levels, α accounts for the efficiency of the electro-optic modulator and p is the pump parameter. The rest of the parameters are related to the external periodic forcing defined above.

By increasing the amplitude of the external forcing, the system undergoes a sequence of subharmonic bifurcations, and for $\beta < 0.1$ the dynamics is restricted to a certain region of the phase space, say $|x_1| < 0.013$, as shown in Figure 5.2. Further increase of β induces an interior crisis, and the clear attractor expansion that can be observed in Figure 5.2 for $\beta \approx 0.1$. This leads to the occurrence of a regime where there is an intermittency between orbits contained in the pre-crisis bounding region and excursions out of it, of period three and four. The set of parameters used in the numerical simulations are $k = 30$, $\alpha = 4$, $\gamma_1 = 10.0643$, $g = 0.05$, $p = 0.01987$, $\gamma_2 = 1.0643$, $z = 10$, $f_0 = 1/7$ and $b_0 = 0.1794$. The stability analysis provides a value of the relaxation oscillation frequency of 0.07, which is around half the frequency of the forcing signal.

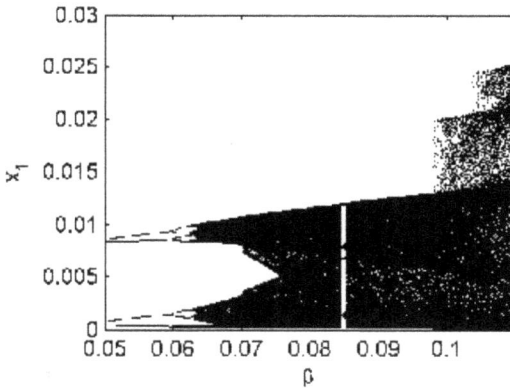

Figure 5.2. Numerical bifurcation diagram for β. Two interior crises are observed, but we are going to study the effect of harmonic perturbations on the laser around the first crisis.

The *phase control scheme* is here implemented as follows. We choose again to perturb harmonically one of the parameters of the system, b_0, because it is easily accessible in the experimental setup. The perturbed parameter becomes a periodic function $b(t) = b_0(1 + \epsilon \sin(2\pi f_0 r t + \phi))$. The phase control scheme relies in an appropriate use of the phase ϕ, once ϵ and r are fixed.

Due to the fact that $F(t)$ depends linearly on the bias b_0, one can clearly see that adding a harmonic perturbation to the bias is equivalent to adding a second periodic forcing, which is one of the possible implementations of the phase control scheme [Qu *et al.* (1995); Yang *et al.* (1996)], corresponding to Eq. (5.3). Thus, for the perturbed system, the forcing term of Eq. (5.5) should read

$$F(t) = \beta \sin(2\pi f_0 t) + \epsilon' \beta \sin(2\pi f_0 r t + \phi) + b_0, \qquad (5.6)$$

with $\epsilon' = b_0 \epsilon / \beta$. We consider ϵ' instead of ϵ, since it enables us to quantify the strength of the applied perturbation in terms of the main periodic forcing.

5.2.3 *Phase control of the laser in the pre-crisis regime*

We consider the role of the phase when the unperturbed laser is placed in the situation previous to the interior crisis, so that no intermittency takes place (not even induced by noise, since we choose to be quite far from the interior crisis). In this case, we characterize the effect of ϕ for fixed values of ϵ' and r by taking records of very long time series where ϕ is slowly varied $\phi \mapsto \phi(t) = 2\pi \mu t$ where $\mu \ll 1/f_0$, i.e. the phase varies very slowly compared to the typical time scale of the laser. Thus, for $t = 0$ the phase difference is $\phi = 0$ and it increases until $t = 1/\mu$, where it is $\phi = 2\pi$. The dynamical state of the system at a certain time t' corresponds essentially to the expected behavior for $\phi = 2\pi \mu t'$.

Let us first analyze the case in which the frequency of the perturbation is the same as the frequency of the main driving, that is, $r = 1$. The experimental long time series for this case is plotted in Figure 5.3. We can observe how there is an increase of the amplitude of the peaks when ϕ is close to 0 and 2π, and a depression as ϕ goes to π. This phenomenon has a simple explanation, indeed

$$F(t) = \beta \sin(2\pi f_0 t) + \epsilon' \beta \sin(2\pi f_0 t + \phi) = \beta' \sin(2\pi f_0 t + \phi_0), \qquad (5.7)$$

where

$$\beta' = \beta \sqrt{1 + \epsilon'^2 + 2\epsilon' \cos \phi}. \qquad (5.8)$$

Notice that we basically have a single forcing, so the resulting ϕ_0 plays an irrelevant role. However, the effective amplitude of the perturbation,

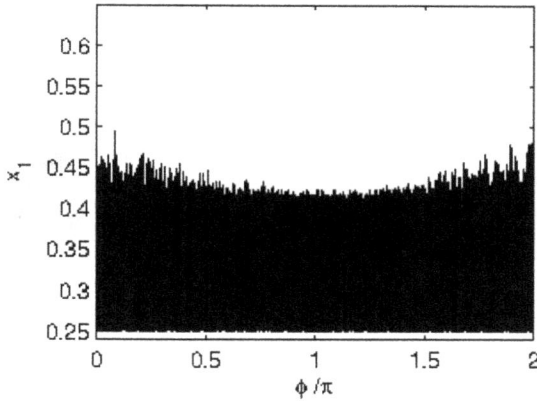

Figure 5.3. Long time series varying the phase ϕ for $r = 1$ $\epsilon' = 0.1$. Note that for $\phi = 0$ and $\phi = 2\pi$ the maxima of the series are increased, as expected.

β', depends on ϕ. Thus, by choosing $\phi = 0$, the effective amplitude of the periodic forcing is increased to a value closer to the critical value so the height of the peaks becomes bigger. Instead, by choosing $\phi = \pi$, β' becomes smaller, so the system is further away from the crisis, and the height of the peaks becomes smaller. We shall point out, in this figure, that with a perturbation of about 10% of the main forcing, the system is not led to the intermittent regime. This is not very relevant by itself, because Eq. (5.8) shows that the necessary amplitude of the perturbation to lead the system to the intermittent regime could be reduced just by placing the unperturbed system closer to the crisis. However, it is an important reference to evaluate the effectiveness of perturbations with different frequencies.

Now we consider the laser in the same unperturbed situation before the crisis and we apply a perturbation whose frequency is the same as the frequency of the unstable periodic orbit involved in the interior crisis [Meucci *et al.* (2005)], that is, $f_0/3$. The two main behaviors observed experimentally are summarized in the two diagrams shown in Figure 5.4. We observe an evident $2\pi/3$ symmetry of the first diagram, which could be deduced from the invariance of Eq. (5.6) under the transformation $t \mapsto t + k/f_0$ and $\phi \mapsto \phi + 2\pi r k$, with k an integer. Figure 5.4a shows the crucial role played by the phase difference. For the same values of the perturbation amplitude, by adjusting the phase, the system can be placed either in an intermittent regime or in the pre-crisis regime. It is important to note that we observe experimentally this significant effect even if the amplitude of the perturba-

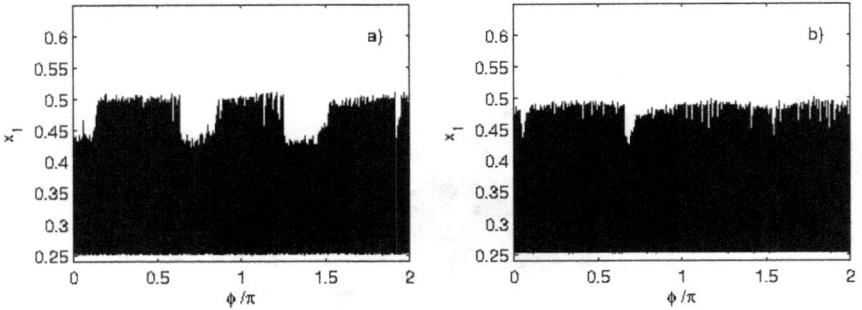

Figure 5.4.   Long time series varying the phase ϕ for $r = 1/3$, (a) $\epsilon' = 0.003$ and (b) $\epsilon' = 0.006$. The diagrams present the expected $2\pi/3$ symmetry, even if in (b) the intervals of ϕ leading to intermittency have merged and the behavior is nearly phase independent.

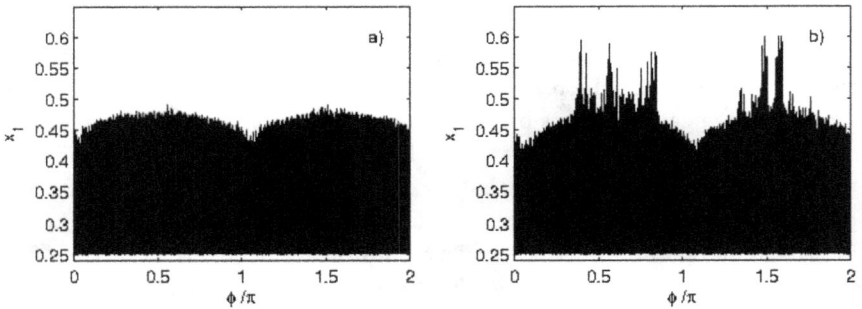

Figure 5.5.   Long time series varying the phase ϕ for $r = 1/2$, (a) $\epsilon' = 0.006$ and (b) $\epsilon' = 0.01$. We have a π symmetry, as expected. The dependence on the phase ϕ is clear, and we observe that in (b) a correct selection of the phase determines whether there is intermittency or not.

tion applied is about 0.3% of the amplitude of the main forcing, which is much smaller than in the $r = 1$ case. However, as we observe in Figure 5.4b for $\epsilon' = 0.006$, there is intermittency for nearly all values of ϕ. In summary, the $r = 1/3$ perturbation is much more effective than the $r = 1$ perturbation to control the intermittency. New features arise when the system is perturbed with the frequency corresponding to the period doubling bifurcation of the system, $f_0/2$. Again, two experimental diagrams are presented to see the effect of the phase, shown in Figure 5.5. We can observe the expected π symmetry in ϕ. On the other hand, Figure 5.5a shows again that the phase difference modulates the maximum height of the peaks, but intermittency does not take place. However, when the perturbation is increased to $\epsilon' = 0.01$, Figure 5.5b, just 1% of the main forcing, the effect of the phase is even clearer: again, the phase enables us to place the system either in the intermittent regime or in the small chaos regime. In the intermittent regime observed in Figure 5.5b, the high amplitude orbits are related with the second interior crisis shown in Figure 5.2, thus a variety of dynamical behaviors is accessible by varying ϕ.

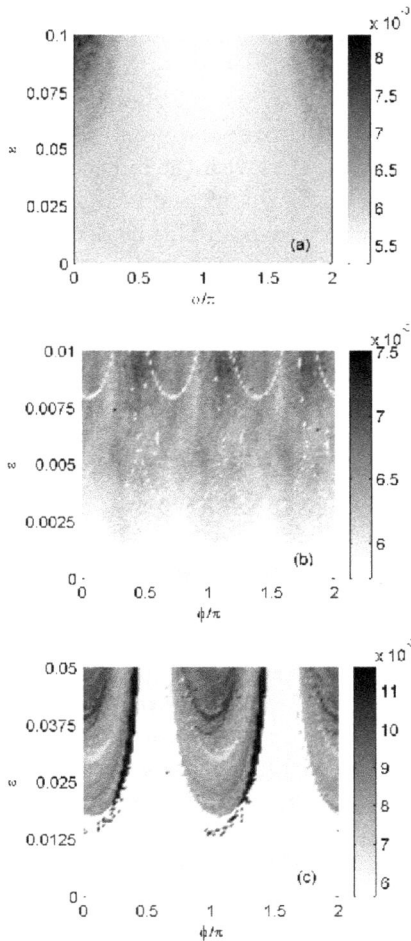

Figure 5.6.   The medium height of the maxima of x_1, $<H>$, as a function of ϵ and ϕ for (a) $r = 1$, (b) $r = 1/3$ and (c) $r = 1/2$.

Numerical calculations provide a confirmation of these results, together
with a deeper insight on the role of the phase. A good indicator to discriminate between the different dynamical states of the laser for different values
of the parameters is

$$< H > = < \max(x_1(t)) >|_{x_1(t) > x_0} \qquad (5.9)$$

where $< \cdot >$ indicates the average over a long time series, and max( ) indicates the relative maximum of the series. The value of x_0 is chosen
in such a way that $< H >$ enables us to distinguish between the small
chaos and the intermittent regime. In the numerical simulations we have
observed that taking $x_0 = 10^{-5}$, that is, neglecting only the extremely
small peaks of the signal is sufficient for this discrimination. We have observed that $< H > \le 0.006$ corresponds to the pre-crisis chaotic regime,
$0.006 < < H > \le 0.0074$ matches with the intermittent regime observed after the first crisis shown in Figure 5.2 and $< H > > 0.0074$ corresponds to
the regime in which there are high amplitude orbits, like those observed in
Figure 5.2 after the second crisis. $< H >$ can be easily computed by numerical integration of the equations of the laser. We study the dependence of
the global dynamics on the parameters of the system by calculating $< H >$
as a function of ϵ and ϕ, fixing r.

Numerical calculations are presented in Figure 5.6. As for the experimental results, we include the calculations for the trivial case $r = 1$ for the
sake of clarity. In this case, Figure 5.6a, the color of the diagram and thus
$< H >$ change smoothly as the parameters vary, from a minimum at $\phi = \pi$
to a maximum at $\phi = 0.2\pi$, as observed in the experiment (Figure 5.3).

For the $r = 1/3$ case, Figure 5.6b, $< H >$ presents the expected $2\pi/3$
symmetry. On the other hand, it can be clearly observed how the value of
$< H >$ increases gradually with ϵ. For a narrow interval of ϵ, approximately
$\epsilon \in [0.002, 0.003]$, depending on ϕ we have values of $< H >$ bigger than 0.006
intercalated with values of $< H >$ smaller than 0.006. This agrees with the
phase-induced transitions between the intermittency and the small chaos
regime observed experimentally. However, as in the experiment, we can see
that if the perturbation amplitude ϵ is further increased with the intervals
of ϕ giving rise to intermittency merge, so intermittency is observed almost
independently of the phase.

Let us finally comment on the results for the $r = 1/2$ case. For small
values of ϵ, $\epsilon < 0.02$, the $< H >$ remains around $< H > \approx 0.005$ almost
independently of the phase. However, when ϵ becomes bigger than a certain
critical value $\epsilon_0 \approx 0.02$, there is a sudden change in the medium height of the

peaks. This sudden transition to a high $< H >$ regime, which corresponds to the dynamical state observed after the second crisis of the laser, is evident from the drastic change of color that we can observe in the diagram of Figure 5.6c, which is fully consistent with the experiments on the laser. Thus, once again we see the important role played by ϕ in placing the system before or after the interior crisis.

5.2.4 *Phase control of the intermittency after the crisis*

Up to now we have shown that the intermittency of the CO_2 laser in the pre-crisis regime can be controlled, by just varying the phase ϕ. In this section, in analogy with [Meucci *et al.* (2005)], we show that the phase control does also work when the unperturbed laser is placed in the post-crisis region.

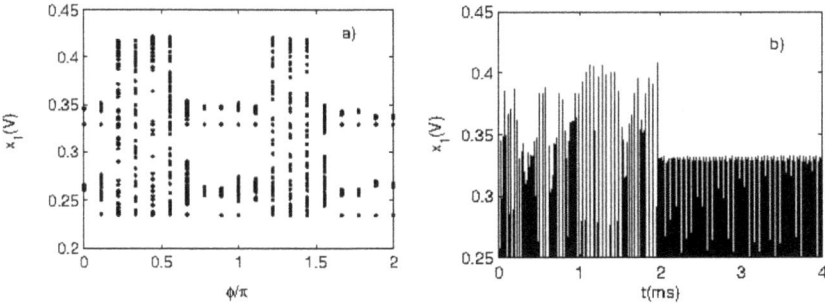

Figure 5.7. Experimental bifurcation diagram (a), showing how an appropriate selection of the phase ϕ can take the laser from an intermittent regime to a small chaos regime. A controlled time series of the laser, where control is applied at $t \approx 2$ ms and the intermittent behavior is nearly immediately suppressed (b).

In order to characterize the role of ϕ in this case we have opted to perform a bifurcation diagram by localizing the maxima of different time series of the laser with different values of ϕ, for $\epsilon = 0.01$ and $r = 1/2$, as shown in Figure 5.7a. We can clearly appreciate a π symmetry in the diagram as in the previous section. We can see how a variation of ϕ allows us to move from the intermittent regime to the small chaos regime. The action of the applied perturbation on the laser is illustrated by Figure 5.7b, where we can see how, once the perturbation is applied (for $t \approx 2$ ms), the system passes from an intermittent regime, with the characteristic large spikes, to a small chaos regime.

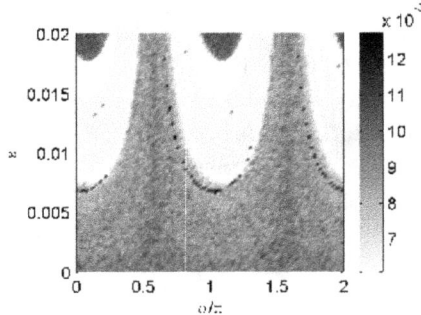

Figure 5.8.   The medium height of the maxima of x_1, $<H>$, as a function of ϵ and ϕ for $r = 1/2$.

We have performed a numerical analysis to see this phenomenon in more detail. We characterize the role of the phase ϕ by calculating $<H>$, as defined in Eq. (5.9), and the results are shown in Figure 5.8. Again, the symmetry induced by our selection of r and the nontrivial role played by the phase ϕ are evident. Thus, we have shown that we can use ϕ to control the intermittency after the interior crisis.

5.3   Conclusions and discussions

In this chapter we have made a thorough exploration of the applications of the phase control technique. Firstly, we have focused on its applications to control chaos in the paradigmatic Duffing oscillator. We have performed numerical simulations confirming the most important properties of this method: that only a correct choice of ϕ can lead the system to a periodic orbit and that, by adequately selecting the phase, the necessary amplitude to suppress chaos can be minimized. By using extensive exploration of parameter space in search of zones of chaos suppression for different values of ϵ and ϕ, we have detected some interesting patterns. Most of the interesting patterns found numerically have been recovered in an experiment with an electronic circuit that mimics the dynamics of the Duffing oscillator with a slight potential asymmetry, even in the presence of noise. This fact suggests that phase control of chaos is robust even in the presence of distortions of the potential symmetry and that all the properties are of quiet general nature, so they must be taken into account when applying this control method to the most diverse dynamical systems.

On the other hand, we have shown that the phase control scheme is able to control the intermittency in a chaotic system close to an interior crisis. First, we have shown both experimentally and numerically how, if we apply a harmonic perturbation to a chaotic CO_2 laser, it is possible to control the crisis-induced intermittency by accurately choosing ϕ. We have seen that this scheme is more effective when the frequency of the perturbation is equal to either the frequency of the unstable periodic orbit involved in the crisis or the frequency involved in the period doubling bifurcation, which can be obtained from experimental time series by using the Fourier transform. Our analysis shows that the application of a periodic modulation to a system close to an interior crisis perturbs its geometry, and such perturbation depends strongly on ϕ, which becomes a key parameter for the global dynamics.

5.4 References

Ott E., Grebogi C. & Yorke J. (1990). Control chaos, *Phys. Rev. Lett.*, 64(11), pp. 1196-1199.

Shinbrot T., Grebogi C., Ott E. & Yorke J.A. (1993). Using small perturbations to control chaos, *Nature*, 363, pp. 411-417.

Aguirre J., d'Ovidio F. & Sanjuán M.A.F. (2004). Controlling chaotic transients: Yorke's game of survival, *Physical Review E*, 69, 016203.

Boccaletti S., Grebogi C., Lai Y.C., Mancini H. & Maza D. (2000). Control of chaos: Theory and applications, *Phys. Rep.*, 329(3), pp. 103-197.

Lima R. & Pettini M. (1990). Suppression of chaos by resonant parametric perturbation, *Phys. Rev. A*, 41(2), pp. 726-733.

Meucci R., Gadomski W., Ciofini M. & Arecchi F.T. (1994). Experimental control of chaos by means of weak parametric perturbations, *Physical Review E*, 49(4), R2528-R2531.

Zambrano S., Brugioni S., Allaria E., Leyva I., Sanjuán M.A.F., Meucci R. & Arecchi F.T. (2006). Numerical and experimental exploration of phase control of chaos, *Chaos*, 16(1), 013111.

Zambrano S., Marino I.P., Salvadori F., Meucci R., Sanjuán M.A.F. & Arecchi F.T. (2006). Phase control of the intermittency in dynamical systems, *Phys. Rev. E*, 74, 016202.

Seoane J.M., Zambrano S., Euzzor S., Meucci R., Arecchi F.T. & Sanjuán M.A.F. (2008). Avoiding escapes in open dynamical systems using phase control, *Phys. Rev. E*, 78(1), 016205.

Qu Z., Hu G., Yang G. & Qin G. (1995). Phase effect in taming nonautonomous chaos by weak harmonic perturbations, *Phys. Rev. Lett.*, 74, pp. 1736-1739.

Yang J., Qu Z. & Hu G. (1996). Duffing equation with two periodic forcings: The phase effect, *Phys. Rev. E*, 53, pp. 4402-4413.

Grebogi C., Ott E. & Yorke, J. (1982). Chaotic attractors in crisis, *Phys. Rev. Lett.*, 48(22), pp. 1507-1510.

Grebogi C., Ott E., Romeiras F. & Yorke J. (1987). Critical exponents for crisis-induced intermittency, *Phys. Rev. A*, 36(11), pp. 5365-5380.

Meucci R., Allaria E., Salvadori F. & Arecchi F.T. (2005). Attractor selection in chaotic dynamics, *Phys. Rev. Lett.*, 95(18), 184101.

Meucci R., Cinotti D., Allaria E., Billings L., Triandaf I., Morgan D. & Schwartz I.B. (2004). Global manifold control in a driven laser: Sustaining chaos and regular dynamics, *Physica D*, 189(1-2), pp. 70-80.

Marino I.P. & Míguez J. (2006). An approximate gradient-descent method for joint parameter estimation and synchronization of coupled chaotic systems, *Physics Letters A*, 351(4-5), pp. 262-267.

Chapter 6

The Jerk Dynamics of Laser's Minimal Universal Model

6.1 Introduction

The model of laser with feedback and minimal nonlinearity belongs to the class of three-dimensional paradigmatic nonlinear oscillator models giving rise to chaos. The minimal universal model for chaos in laser with feedback contains three key nonlinearities, two of which are of the type xy, where x and y are the fast and slow variables. The third one is of the type xz^2, where z is an intermediate feedback variable. In a previous publication, Meucci *et al.* [2021] have analytically demonstrated that it is essential to produce chaos via local or global homoclinic bifurcations. Its electronic implementation in the range of kilo Hertz region has also confirmed its potential to describe phenomena evolving on different time scales.

During these last two decades, the seminal works of Gottlieb [1996] and Sprott [2003, 2011] have triggered an increasing interest in the study of chaotic oscillators based on jerk equations, that is, oscillators which can be completely described by third-order ordinary differential equations of the form $\dddot{x} = f(\ddot{x}, \dot{x}, x)$. Recently, Buscarino *et al.* [2014] and Xu and Cao [2020] have provided the jerk forms dynamics of Chua's circuit.

In this paper, following the method proposed by Buscarino *et al.* [2014], we prove (to our knowledge for the first time) that at least two jerk forms of laser's minimal universal model can be derived: the two forms refer respectively to the variables z and y of laser's minimal universal model. Although a jerk equation in terms of the variable x cannot be provided by such methods, it may be possible to obtain it with the *controllable canonical form* used by Xu and Cao [2020]. Let us note that due to the quadratic term in xz^2 in the original three-order laser's minimal universal model, the jerk form in y is bi-valuated, i.e. depends on the square root of z which can be either positive or negative. Such peculiarity makes the analysis very difficult and prevents from plotting its attractor. That is the reason why we have focused our analysis on the first jerk form in z. Then, by making a comparison of the Lyapunov Characteristic Exponents, eigenvalues and attractors between the original three-order laser's minimal universal model and its first jerk form in z, we have demonstrated the *topological equivalence* of both systems.

The chapter is organized as follows. In the next section, some introductive materials on jerk equations and laser's minimal universal model are reported. Then, in Section 6.3 the jerk equations of laser's minimal universal model are derived. Mathematical and numerical results concerning the stability analysis are reported in Section 6.4. The electronic realization of the jerk laser's minimal universal model is proposed in Section 6.5.

6.2   Preliminaries

At the end of the seventies, Stephen H. Schot [1978] published a very interesting paper in which he made a short history of the origin of the use of the concept of jerk, i.e. the time rate of change of the acceleration. He explained that:

> "The French geometer Transon in 1845 was probably the first to consider the third time derivative of distance in mechanics and he uses the term virtualité for it. Transon computes the normal component of the jerk and expresses it in terms of what is now called the aberrancy (he uses the term déviation de courbure). Other early writers who treat the third derivative use of neutral terms such as second acceleration or higher acceleration for it. Thus, Resal resolves the jerk along a space curve into tangential, normal, and binormal components and Somoff first establishes recursion formulas for these components of the higher-order accelerations in terms of those of the ordinary acceleration. Subsequently, the term jerk for the second acceleration seems to have gradually entered the literature of physics and without any explicit rational explanation for its use."

According to Pr. Christian Mira (personal communication), at the end of the nineteenth century, the French scientist Martin Haag (not to be confused with Jules Haag) made use of the concept of "over-acceleration" or jerk in a mathematical analysis [Haag, 1879]. Then, as recalled by Schot [1978]:

> "An unequivocal definition of the jerk as "the derivative of acceleration with respect to time" was given in 1928 by Melchior who justifies the use of the term by referring to the physiological sensation experienced by large changes in acceleration ... The term is now commonly used in mechanics and is being adopted in other areas of physics as well."

More than half a century later, Julian Clinton Sprott [1994] found probably the simplest three-dimensional first-order system for chaos. Two years after, Hans Gottlieb [1996] wondered "what is the simplest jerk function that gives chaos?" where the jerk function is of the form: $\dddot{x} = f(x, \dot{x}, \ddot{x})$. According to Buscarino *et al.* [2014]: "If a mechanical interpretation of the variable x in terms of displacement is given, the jerk equation can be viewed as an equation where the derivative of the acceleration is involved, that is, where a measure of the instantaneous variation of the acceleration is included." Since the question of the simplest jerk function is still open, in this work, we will only focus on the *topological equivalence* between the original model and its corresponding jerk equations. To this aim, we will use the Lyapunov Characteristic Exponents, the bifurcation diagram and the shape of the attractor in the phase space to show that both laser's minimal universal model and its jerk equations in z are topologically equivalent. In general, given a nonlinear system of order n and considered one of its n state variables, say variable x_i, there is no guarantee that it can be rewritten in the equivalent form:

$$\frac{d^n x_i}{dt^n} = f\left(\frac{d^{(n-1)} x_i}{dt^{n-1}}, \frac{d^{(n-2)} x_i}{dt^{n-2}}, \ldots, x_i\right). \tag{6.1}$$

Although Xu and Cao [2020] have proposed a scheme to implement the Jerk form of the Chua system family using a *controllable canonical form* applied in linear systems, it does not seem (to our knowledge) that this scheme has been used for more general system such as laser's minimal universal model. So, we will follow in this work the method used by Buscarino *et al.* [2014]. Thus, when it is possible to derive a jerk equation, the state space normal form of the system can also be written as follows:

$$\frac{d\tilde{x}_1}{dt} = \tilde{x}_2,$$

$$\frac{d\tilde{x}_2}{dt} = \tilde{x}_3,$$

$$\vdots$$

$$\frac{d\tilde{x}_n}{dt} = f\left(\tilde{x}_1, \tilde{x}_2, \ldots, \tilde{x}_{n-1}\right), \qquad (6.2)$$

with $\tilde{x}_1 = x_i$.

The minimal universal model for chaos in laser reads [Meucci *et al.* (2021)]:

$$\frac{dx}{dt} = -\epsilon_1 x \left(1 + k_1 z^2 - p_0 y\right),$$

$$\frac{dy}{dt} = -y - xy + 1, \qquad (6.3)$$

$$\frac{dz}{dt} = -\epsilon_2 \left(z - B_0 + B_1 x\right),$$

Let us note that with the parameter set used in our experiment and analysis, $\epsilon_1 \gg 1$ and $\epsilon_2 \gg 1$. So, model (6.3) is a *slow–fast* dynamical system involving two *fast* times scales. In the following we will use the parameter set in our experiment and analysis:

$$\epsilon_1 = 200 \ , \ \epsilon_2 = 6 \ , \ k_1 = 12 \ , \ p_0 = 1.208 \ , \ B_1 = 0.555.$$

and B_0 $(0.12 < B_0 < 0.125)$ will play the role of a control parameter.

6.3   The jerk form of laser's minimal universal model

Let us consider Eq. (6.3). Is it possible to obtain the following jerk forms?

$$\dddot{x} = f_1\left(x, \dot{x}, \ddot{x}\right), \qquad (6.4)$$

$$\dddot{y} = f_2\left(y, \dot{y}, \ddot{y}\right), \qquad (6.5)$$

$$\dddot{z} = f_3\left(z, \dot{z}, \ddot{z}\right). \qquad (6.6)$$

We show here that this question has a positive answer only for the forms (6.5) and (6.6). Nevertheless, due to the quadratic term in xz^2 in the original three-order laser's minimal universal model, the jerk form in y is bi-valuated, i.e. depends on the square root of z which can be either positive or negative. This prevents from plotting its attractor and precludes any analysis of the jerk form (6.5), i.e. the $2 - 1 - 3$ structure. That is the reason why, in the following, we will focus our analysis on the first jerk form in z (6.6), i.e. the $3 - 2 - 1$ structure.

6.3.1 *Jerk form in z*

Starting from the first equation of (6.3), we obtain:

$$y = \frac{1}{\varepsilon_1 p_0} \left(\frac{\dot{x}}{x} + \varepsilon_1 + \varepsilon_1 k_1 z^2 \right). \tag{6.7}$$

From the third equation of (6.3), we deduce that:

$$x = \frac{1}{\varepsilon_2 B_1} \left(\varepsilon_2 B_0 - \dot{z} - \varepsilon_2 z \right). \tag{6.8}$$

Taking the time derivative of this Eq. (6.8) leads to:

$$\dot{x} = \frac{1}{\varepsilon_2 B_1} \left(-\ddot{z} - \varepsilon_2 \dot{z} \right). \tag{6.9}$$

By using the expression of $\dot{x}$ and x, i.e. Eqs. (6.8)–(6.9) to obtain the ratio $\dot{x}/x$ and by replacing in Eq. (6.7), we find that:

$$y = \frac{1}{\varepsilon_1 p_0} \left(\frac{\ddot{z} + \varepsilon_2 \dot{z}}{\dot{z} + \varepsilon_2 z - \varepsilon_2 B_0} + \varepsilon_1 + \varepsilon_1 k_1 z^2 \right). \tag{6.10}$$

The time derivative of Eq. (6.10) reads:

$$\dot{y} = \frac{1}{\varepsilon_1 p_0} \left(\frac{(\dddot{z} + \varepsilon_2 \ddot{z})(\dot{z} + \varepsilon_2 z - \varepsilon_2 B_0) - (\ddot{z} + \varepsilon_2 \dot{z})^2}{(\dot{z} + \varepsilon_2 z - \varepsilon_2 B_0)^2} + 2\varepsilon_1 k_1 z \dot{z} \right). \tag{6.11}$$

Then, from the second equation of (6.3), we have:

$$\dot{y} + y(1 + x) = 1. \tag{6.12}$$

By replacing x in Eq. (6.12) by the expression (6.8) and $\dot{y}$ by the expression (6.11) and solving the resulting equation with respect to $\dddot{z}$, we obtain:

$$\dddot{z} = \varepsilon_1 p_0 \left(\dot{z} + \varepsilon_2 z - B_0 \varepsilon_2\right) \left[1 - \frac{2k_1 z \dot{z}}{p_0} - \frac{\varepsilon_2 \ddot{z}}{\varepsilon_1 p_0 \left(\dot{z} + \varepsilon_2 z - B_0 \varepsilon_2\right)} + \frac{\left(\ddot{z} + \varepsilon_2 \dot{z}\right)^2}{\varepsilon_1 p_0 \left(\dot{z} + \varepsilon_2 z - B_0 \varepsilon_2\right)^2}\right.$$

$$\left. - \frac{\varepsilon_2 B_0 + \varepsilon_2 B_1 - \dot{z} - \varepsilon_2 z}{\varepsilon_1 \varepsilon_2 p_0 B_1} \left(\varepsilon_1 + \varepsilon_1 k_1 z^2 + \frac{\ddot{z} + \varepsilon_2 \dot{z}}{\dot{z} + \varepsilon_2 z - B_0 \varepsilon_2}\right)\right] \tag{6.13}$$

Then, the jerk form in z (6.6) is obtained by posing:

$$\dot{z} = y, \quad \ddot{z} = x, \quad \dddot{z} = f_3\left(z, \dot{z}, \ddot{z}\right) = x. \tag{6.14}$$

Considering Eq. (6.3) of laser's minimal universal model, we obtain the dynamics of the jerk system:

$$\frac{dx}{dt} = f_3\left(x, y, z\right),$$

$$\frac{dy}{dt} = x, \tag{6.15}$$

$$\frac{dz}{dt} = y,$$

where

$$f_3\left(x, y, z\right) = \varepsilon_1 p_0 \left(y + \varepsilon_2 z - B_0 \varepsilon_2\right) \left[1 - \frac{2k_1 z y}{p_0} - \frac{\varepsilon_2 x}{\varepsilon_1 p_0 \left(y + \varepsilon_2 z - B_0 \varepsilon_2\right)} + \frac{\left(x + \varepsilon_2 y\right)^2}{\varepsilon_1 p_0 \left(y + \varepsilon_2 z - B_0 \varepsilon_2\right)^2}\right.$$

$$\left. - \frac{\varepsilon_2 B_0 + \varepsilon_2 B_1 - y - \varepsilon_2 z}{\varepsilon_1 \varepsilon_2 p_0 B_1} \left(\varepsilon_1 + \varepsilon_1 k_1 z^2 + \frac{x + \varepsilon_2 y}{y + \varepsilon_2 z - B_0 \varepsilon_2}\right)\right] \tag{6.16}$$

Equations (6.15) provide in different state space a representation of system (6.3) and thus maintain its structural properties. The three-dimensional attractors for the previously defined parameters and for $B_0 = 0.1246$ are reported in the following figures. The original laser's minimal universal model chaotic attractor is reported in Figure 6.1a, the attractor of the equivalent jerk system represented by Eqs. (6.15) is reported in Figure 6.1b.

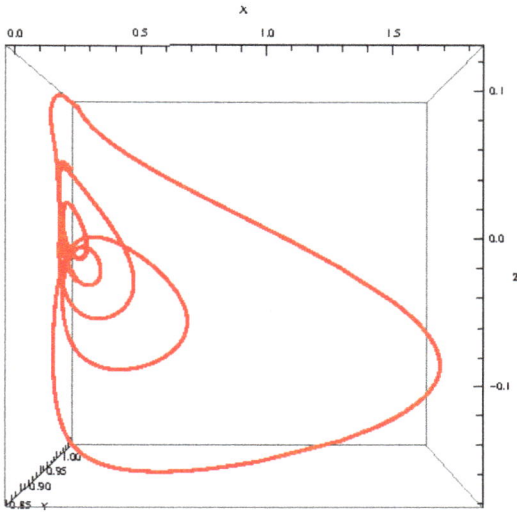

(a)

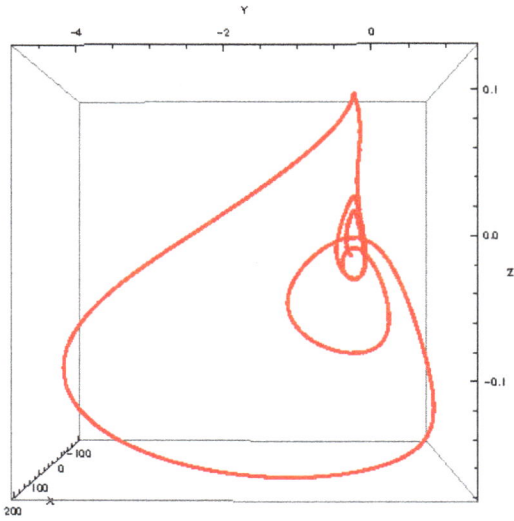

(b)

Figure 6.1.   Phase portraits of (a) the original laser's minimal universal model (6.3) and (b) its jerk form in z (6.15).

In order to state the topological equivalence between the original representation of the original laser's minimal universal model (6.3) and its jerk form in z (6.15), we have performed a stability analysis including the fixed points stability, the occurrence of Hopf bifurcation, the representation of the bifurcation diagram and the computation of the Lyapunov Characteristic Exponents for the jerk form in z (6.15) that we have compared to the stability analysis of the original laser's minimal universal model (6.3).

6.3.2 *Stability analysis*

By using the classical nullclines method, it can be shown that the dynamical system (6.15) admits four positive fixed points.

$$I_1 (0, 0, B_0), \quad I_{2,3,4} (0, 0, z^*), \tag{6.17}$$

where the expression of z^* (too large to be explicitly written here) is the solution of the following cubic polynomial and depends on the control parameter B_0.

$$k_1 z^3 - k_1 (B_0 + B_1) z^2 + z + B_1 (p_0 - 1) - B_0 \tag{6.18}$$

If we replace z^* by 0 in the above equation (6.18) we obtain: $B_1 (p_0 - 1) - B_0 < 0$. Now, if we replace z^* by B_1, we find $B_0(p_0 - k_1 B_1^2 - 1) > 0$. This means that:

$$B_0 - B_1 (p_0 - 1) < z^* < B_1$$

Let us note that the left part of this inequality exactly corresponds to the one found in our previous publication [Meucci *et al.* (2021)].

6.3.3 *Jacobian matrix*

The Jacobian matrix of the dynamical system (6.15) reads:

$$J = \begin{pmatrix} \dfrac{\partial f_3}{\partial x} & \dfrac{\partial f_3}{\partial y} & \dfrac{\partial f_3}{\partial z} \\ 1 & 0 & 0 \\ 0 & 1 & 0 \end{pmatrix} \tag{6.19}$$

By replacing the coordinate of the fixed points I_1 (6.17) in the Jacobian matrix (6.18) one obtains the Cayley–Hamilton third degree eigenpolynomial which reads:

$$(\lambda + \epsilon_2) \left[\lambda^2 + \lambda + \epsilon_1 \left(1 + B_0^2 k_1 - p_0 \right) \right] = 0 \tag{6.20}$$

It follows that the first eigenvalue $\lambda_1 = -\epsilon_2 < 0$ and the two others are such that $\lambda_2 \lambda_3 = \epsilon_1 (1 + B_0^2 k_1 - p_0)$. Thus, provided that:

$$p_0 < 1 + B_0^2 k_1, \tag{6.21}$$

the fixed point I_1 is a *saddle-node*. Moreover, such condition (6.21) provides an upper boundary for the control parameter B_0:

$$B_0 < \sqrt{\frac{p_0 - 1}{k_1}}. \tag{6.22}$$

Let us note that if $B_0 = B_1 (p_0 - 1)$, the fixed points $I_{1,2}$ read $I_1 (0, 0, B_1 (p_0 - 1))$, $I_2 (0, 0, 0)$ and the two others read: $I_{3,4} (0, 0, z^* = \frac{1}{2} (B_1 p_0 \pm \sqrt{B_1^2 p_0^2 - 4/k_1}))$. In these conditions and according to Eq. (6.22), it can be stated (exactly as in our previous publication [Meucci *et al.* (2021)]) that:

$$B_1 (p_0 - 1) < B_0 < \sqrt{\frac{p_0 - 1}{k_1}}. \tag{6.23}$$

Thus, for $B_0 = B_1 (p_0 - 1)$, computation of the eigenvalues show that I_1 is still a *saddle-node*, I_3 is a *saddle-focus* while the two other fixed points are stable foci. For $B_0 = \sqrt{(p_0 - 1)/k_1}$, the computation of the eigenvalues show that I_1 is stable while I_2 is a *saddle-node* and I_3 a *saddle-focus*. Such a result will enable to explain the limits of the bifurcation diagram presented below (see Figures 6.2 and 6.3) outside which no attractor can exist. Then, while using the parameter set of our experiment, i.e. for any value of $B_0 \in [B_1 (p_0 - 1), \sqrt{(p_0 - 1)/k_1}]$ it can be shown that I_3 is a *saddle-focus* (the first eigenvalue is real and positive while the two others are complex conjugate with negative real parts). This implies that a Hopf bifurcation occurs within this interval [Hopf (1942); Andronov *et al.* (1971); Marsden & McCracken (1976); Kuznetsov (2004)]. Nevertheless, since the fixed points are solutions of a cubic polynomial, the value of the control parameter B_0 for which such a Hopf bifurcation occurs cannot be given in a simple mathematical way. However, we have numerically computed this Hopf bifurcation parameter value $B_0 \approx 0.12036$. Let us note that this is exactly the same value as the one found in our previous publication [Meucci *et al.* (2021)]. In the following we will use the parameter set in our experiment and analysis:

$$\epsilon_1 = 200 \ , \ \epsilon_2 = 6 \ , \ k_1 = 12 \ , \ p_0 = 1.208 \ , \ B_1 = 0.555.$$

6.3.4 *Bifurcation diagram*

Thus, in order to highlight the effects of the control parameter B_0 changes
on the topology of the attractor we have built the bifurcation diagram of the
jerk form in z of laser's minimal universal model (6.15) (see Figure 6.2b)
that we have compared to the bifurcation diagram of the original laser's
minimal universal model (6.3) (see Figure 6.2a). Both Figures 6.2a and
6.2b clearly demonstrate the equivalence of the two representations.

We have also compared the zoom of such bifurcation diagrams for $B_0 \in$
$[0.12410, 0.124145]$ (see Figures 6.3a and 6.3b).

Here again these bifurcation diagrams are very similar. The same holds
for the time series of the original laser's minimal universal model (6.3) and
its jerk form in z (6.15). In order to confirm the *topological equivalence* between both systems, Lyapunov Characteristic Exponents (LCE) have been
computed for the jerk form in z of the laser's minimal universal model (6.15)
and compared to those obtained for the original laser's minimal universal
model (6.3).

6.3.5 *Numerical computation of the Lyapunov exponents*

As previously, we have used the algorithm developed by Marco Sandri [1996]
for Mathematica® to perform the numerical calculation of the Lyapunov
Characteristics Exponents (LCE) of the original laser's minimal universal
model (6.3). LCEs values have been computed within each considered
interval ($B_0 \in [0.123, 0.1234]$ and $[0.1235, 0.125]$). For 0.1246, Sandri's
algorithm has provided the following LCEs $(+0.2, 0, -7.32)$. By using the
same algorithm we found exactly the same LCEs for the jerk form in z of
the original laser's minimal universal model (6.15).

6.4 Experimental part

Experimental set up used for laser's minimal universal model has been
revisited and now illustrated in Figure 6.4.

The electronic components are the same used in the original configura-
tion but with some adjustments in the resistor values near the conditions
of homoclinic chaos discussed before. Another difference is the use of a
different digital scope (Tektronix) in order to highlight the thickness of the
traces more than its temporal persistence. In Figure 6.5, we can clearly
distinguish traces originated after the rejection mechanism near the saddle

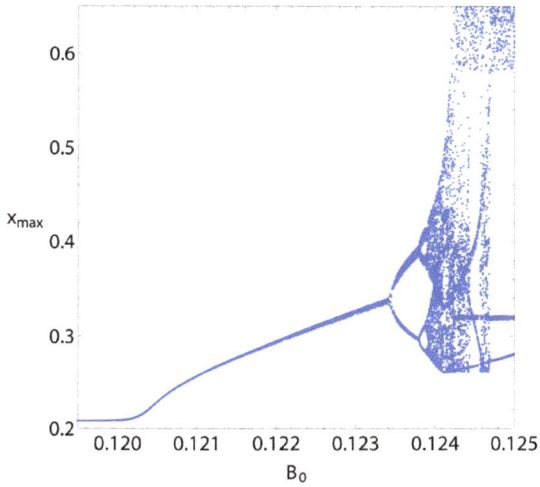

(a)

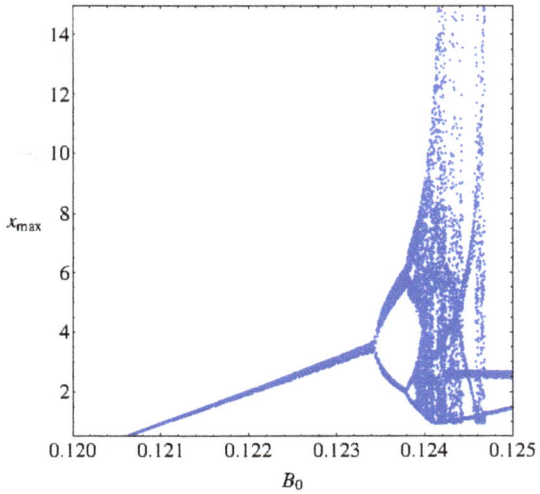

(b)

Figure 6.2. Bifurcation diagrams of (a) the original laser's minimal universal model (6.3) and (b) its jerk form in z (6.15).

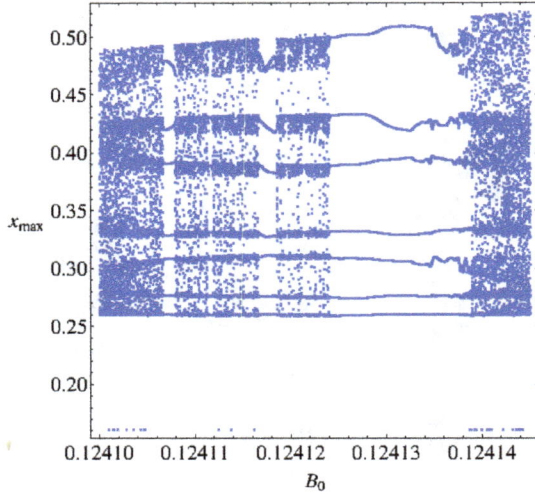

(a)

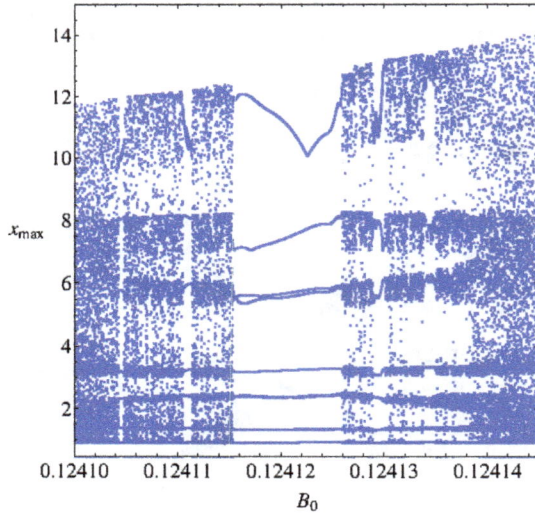

(b)

Figure 6.3.   Zoom of bifurcation diagrams of (a) the original laser's minimal universal model (6.3) and (b) its jerk form in z (6.15).

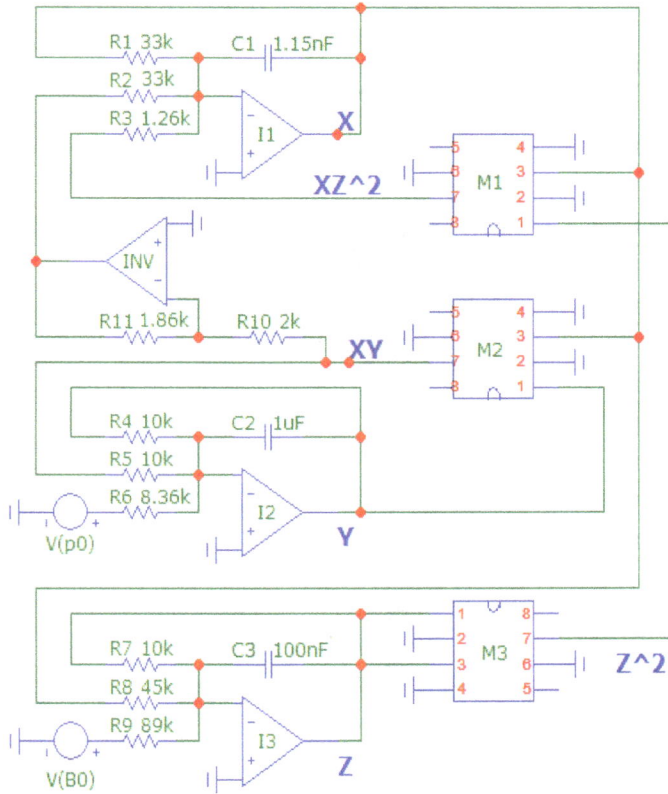

Figure 6.4. Circuit diagram of laser's minimal universal model. M_1, M_2 and M_3 are three analog multipliers (MLT04 by Analog Devices). I_1, I_2 and I_3 are integrators (LT1114 by Analog Devices). *INV* is an inverting gain implemented with the same kind of differential amplifier. $V(p_0)$ and $V(B_0)$ are two adjustable bias voltages.

focus leading to a small chaotic tangle. The chaotic evolution near the saddle focus disappears when the trajectory is attracted by the saddle node with zero intensity solution. This competition mechanism between these two unstable fixed points is responsible for homoclinic chaos with pulses of the same shape but erratically separated in time due to the time spent near the saddle focus. The chaotic fine structure is similar to the one reported in [Pisarchik *et al.* (2000)] where simulations were performed on a more refined laser model named the four-level laser model.

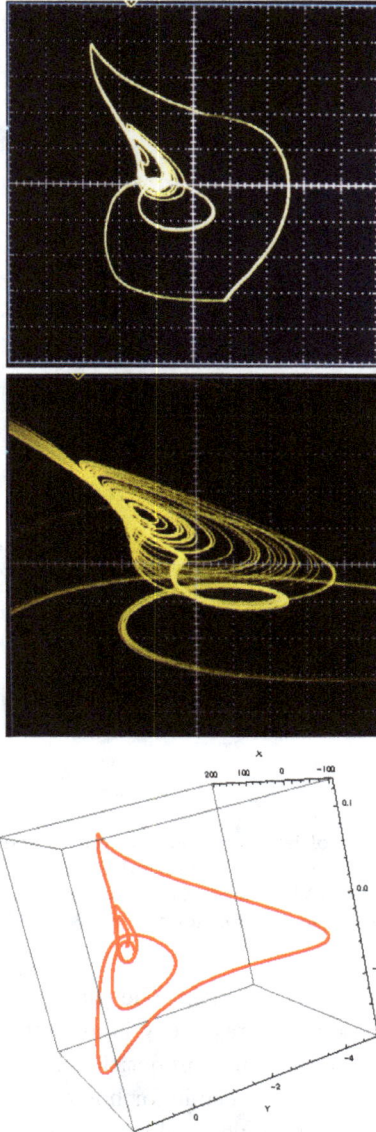

Figure 6.5.   Oscilloscope snapshots of phase portraits of laser's minimal universal model
(6.3) corresponding to (a) $V(p_0) = 3.53V$ and (b) $V(B_0) = 3.64V$ and (c) its jerk form
(6.15) in z. (a) Phase space portrait projected in $x-z$ plane; (b) expansion of the chaotic
tangle region; (c) simulation of the model jerk form.

Considering the limits imposed by analog simulations, the desired dynamics of the oscillator is obtained by fine adjustments of two bias voltages $V(p_0)$ and $V(B_0)$ accounting for the parameters, p_0 and B_0 in Eqs. (6.3) and (6.15). This scheme is exactly the same as the one used in our previous publication [Meucci *et al.* (2021)]. The attractors in the $x - z$ phase space for the $B_0 = 0.1246$ parameter value is reported in Figure 6.5a. The comparison with the attractor obtained by numerical integration of the jerk form in z of the laser's minimal universal model (6.15) (see Figure 6.5c) highlights a very high level of similarity between them.

6.5   Discussion

In this chapter, two jerk forms of the laser's minimal universal model have been derived and the corresponding state space representation in z has been presented. Then, its stability analysis has enabled us to show that the jerk form in z of laser's minimal universal model (6.15) has the same fixed points features (saddle-foci), the same bifurcation diagram, the same Hopf bifurcation parameter value, the same Lyapunov Characteristic Exponents and the same attractor shape. Moreover, we have also observed that an electronic circuit implementing the original laser's minimal universal model (6.3) performed in our previous publication [Meucci *et al.* (2021)] perfectly corresponds to the jerk form in z of the laser's minimal universal model (6.15) (see Figures 6.4a and 6.4b). Thus, the *topological equivalence* between both original laser's minimal universal model (6.3) and its jerk form in z (6.15) is clearly established. In a previous publication [Ginoux *et al.* (2022)], by using the *Flow Curvature Method*, we have been able to state a link between the *curvature* and *energy* of planar generalized Liénard systems. This has been made possible because this kind of system of two ordinary differential equations of order one can be transformed into a nonlinear second order ordinary differential equation. The transformation of systems of three ordinary differential equations of order one into a third order ordinary differential equation, i.e. a jerk equation thus enables to open some promising developments for assessing the energy of nonlinear and chaotic three-dimensional dynamical systems.

6.6   References

Meucci R., Euzzor S., Arecchi F.T. and Ginoux J.M. (2021). Minimal universal model for chaos in laser with feedback, *Int. J. Bifurcation Chaos*, 31(04), 2130013.

Gottlieb H.P.W. (1996). Question 38. What is the simplest jerk function that gives chaos? *Amer. J. Phys.*, 64, 525.

Sprott J.C. (2003). *Chaos and Time-Series Analysis* (Oxford University Press).

Sprott J.C. (2010). *Elegant Chaos: Algebraically Simple Chaotic Flows* (World Scientific, Singapore).

Buscarino A., Fortuna L. and Frasca M. (2014). The jerk dynamics of Chua's circuit, *Int. J. Bifurcation Chaos*, 24(06), 1450085.

Xu W. and Cao N. (2020). Jerk forms dynamics of a Chua's family and their new unified circuit implementation, *IET Circuits Devices Syst.*, 15, pp. 755-771.

Schot H.S. (1978). Jerk: The time rate of change of acceleration, *American Journal of Physics*, 46, pp. 1090-1094.

Haag M. (1879). Note sur les relations entre les éléments caractéristiques d'une courbe gauche et les accélérations du point qui la décrit, *Bulletin de la S. M. F.*, tome 7, pp. 140-143.

Sprott J.C. (1994). Some simple chaotic flows, *Phys. Rev. E*, 50(2), R647-R650.

Hopf E. (1942). Abzweigung einer periodischen Lösung von einer stationären Lösung eines Differentialsystems, *Berichte der MathematischPhysikalischen Klasse der Sächsischen Akademie der Wissenschaften zu Leipzig*, Band XCIV, Sitzung vom 19. January 1942, pp. 3-22. See L.N. Howard and N. Kopell, A Translation of Hopf's Original Paper, pp. 163-193 and Editorial Comments, pp. 194-205 in J. Marsden and M. McCracken.

Andronov A., Leontovich E., Gordon I. & Maier A. (1971). *Theory of Bifurcations of Dynamical Systems on a Plane*, Israel Program for Scientific Translations, Jerusalem.

Marsden J. & McCracken M. (1976). *Hopf Bifurcation and its Applications* (Springer-Verlag, New York).

Kuznetsov Yu. A. (2004). *Elements of Applied Bifurcation Theory* (Springer-Verlag, New York), Third edition.

Sandri M. (1996). Numerical calculation of Lyapunov exponents, *The Mathematica Journal*, 6(3), 78-84.

Pisarchik A.N., Meucci R. & Arecchi F.T. (2000). Discrete homoclinic orbits in a laser with feedback, *Physical Review E*, 62(6), 8823.

Ginoux J.M., Lebiedz D., Meucci R. & Llibre J. (2022). Flow curvature manifold and energy of generalized Liénard systems, *Chaos, Solitons & Fractals*, 161, 112354.

Chapter 7

A Short History of the Discovery of LASER

7.1 The nature of light

Since ancient times, the nature of light has been the subject of much questioning. During the seventeenth and eighteenth centuries, two theories of light were opposed: the corpuscular theory and the wave theory. In his *Hypothesis of Light* of 1675 and then, in his *Opticks* of 1704, Isaac Newton claimed that

"Light was made of very small corpuscles."[1]

For him, light was composed of corpuscles (particles of matter) which were emitted in all directions from a source. During the same period, Christiaan Huygens worked out a mathematical wave theory of light in 1678 and published it in his *Treatise on Light* in 1690. He proposed that light was emitted in all directions as a series of waves in a medium called the *luminiferous æther*. Of course, Newton's reputation helped the particle theory of light to hold sway during the eighteenth century.

In 1802, Thomas Young showed by means of his now famous double-slit experiment that Newton's corpuscular theory was unable to explain the observed phenomenon and so, that light behaved as wave. In fact, by considering that one particle of light passes through the first slit and another by the second one, one would expect to observe light on the screen, since obviously some light plus light may only give light. Nevertheless, Young's diffraction experiment showed that it provides shadow. A result that seemed to invalidate Newton's particle theory of light, at least for a while ...

---

[1] See Newton (1704).

Newton's corpuscular theory implied that velocity of light would be greater in a denser medium like water than in the air, while the wave theory of Huygens implied the opposite. At that time, the speed of light could not be measured accurately enough to decide which theory was correct. The first to make a sufficiently accurate measurement was Léon Foucault, in April 1850. His result supported the wave theory and the classical particle theory was finally abandoned, only to partly re-emerge in the twentieth century.

In 1862, James Clerk Maxwell calculated that the speed of propagation of an electromagnetic field is approximately that of the speed of light. He considered this to be more than just a coincidence, commenting:

> "We can scarcely avoid the inference that light consists in the
> transverse undulations of the same medium which is the cause
> of electric and magnetic phenomena."[2]

7.2   The birth of Quantum Mechanics

In 1900, Max Planck sought to explain the phenomena of *incandescence*, i.e. the emission of electromagnetic radiation (including visible light) from a hot body as a result of its high temperature. In the attempt to bring theory into agreement with experiment, he then showed that light or electromagnetic energy is not exchanged with matter in a *continuous* way but by packets of energy. He thus postulated that this electromagnetic energy is *absorbed* or *emitted* in discrete packets, which he calls quanta. The energy is given by the formula $E = h\nu$ with ν the frequency of the radiation and h called in fact Planck's constant.[3]

In 1905, in order to explain the *photoelectric effect*, i.e. the emission of electrons when electromagnetic radiation, such as light, hits a material, Albert Einstein had the idea to return to the Newton's corpuscular light conception.[4] He then showed that each corpuscle of light has a determined energy that he called *lichtquanta* (*light quantum*). This energy is equal to the product of the frequency ν of the light by the constant h.

In 1913, Niels Bohr presented an atomic model consisting of a small, dense nucleus surrounded by orbiting electrons — similar to the structure

---

[2]See Maxwell (1862).

[3]According to Planck, the letter h is the abbreviation of the German words Hilfsgröße ("auxiliary variable").

[4]See Einstein (1906).

of the Solar System, but with attraction provided by electrostatic forces in place of gravity. In this planetary representation, electrons are distributed in different orbits around the nucleus. The farther the electrons are from the nucleus, the greater their energy. Thus, each orbit corresponds to an energy level. Bohr then showed that electrons have the possibility of passing from one energy level to another by emitting or absorbing a quantum of energy, i.e. a *photon*.[5]

In 1913, only two interaction processes between atoms and radiation were known: *absorption* and *spontaneous emission*.

During *absorption* (see Figure 7.1, left), the atom passes from the *ground state* (lower level 1) with energy E_1 to the *excited state* (upper level 2) with energy E_2 by absorbing a photon. A photon disappeared from the incident wave, and the latter is attenuated.

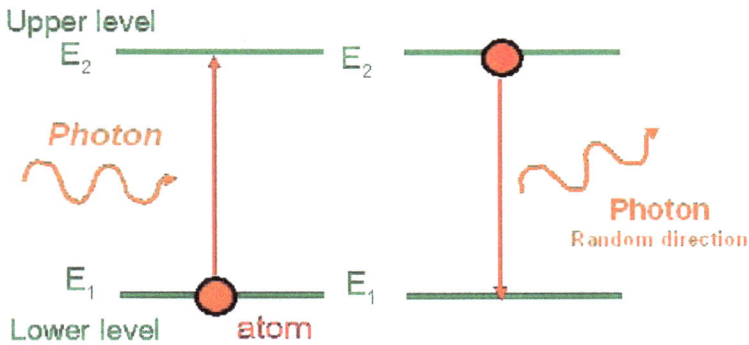

Figure 7.1. *Absorption* and *spontaneous emission* processes.

During *spontaneous emission* (see Figure 7.1, right), the atom initially in the *excited state* with energy E_2 goes back down to the *ground state* with energy E_1 by emitting a photon. This photon is emitted in a random direction after a random time. Upper level 2 is depopulated to the benefit of lower level 1.

In two articles published in 1916 and 1917, Albert Einstein[6] introduced a third process: *stimulated emission*. It is the inverse process of *absorption*, occurring, as in the presence of incident radiation resonant with the transi-

---

[5]The term "photon" was introduced by Gilbert Lewis in 1926 for the smallest unit of radiant energy (light). This name was adopted for what Einstein had called a light quantum (lichtquanta in German).

[6]See Einstein (1916 and 1917).

tion frequency of the atoms. During *stimulated emission* (see Figure 7.2), a photon of energy $h\nu$ induces the de-excitation of an atom from upper level 2 to lower level 1. This is accompanied by the emission of an inductor photon of the same frequency, same direction of propagation and same state of polarization. The incident beam of light is "increased" with identical photons to come and create an amplification of the light. The photon created by stimulated emission has the same properties as the "trigger" photon (frequency, phase, direction of propagation, state of polarization).

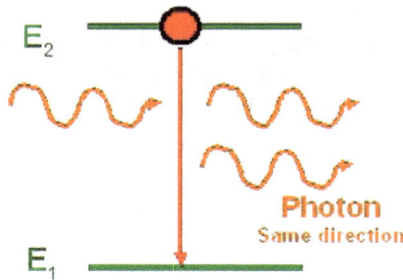

Figure 7.2.   *Stimulated emission* processes.

However, for the *stimulated emission* process to occur, there must be more atoms in the upper level than in the lower one. However, in matter, particles are much more numerous in the ground state than in an excited state. It was then necessary to find a means of reversing the thermal tendency to obtain, in the medium, more particles in an excited state than in a fundamental state, that is to say to carry out a *population inversion*. At that time, this seemed impossible because the coefficient of probability of spontaneous fallout of atoms $A_{2\rightarrow1}$ is very large. Albert Einstein had notably shown that this probability was proportional to the cube of the frequency. As we will see in the next section, *stimulated emission* is the basis of laser functioning.

7.3   From MASER to LASER

In the beginning of the fifties, Charles Townes worked on microwave microscopy of molecules using electronic oscillators. He realized that Einstein's *stimulated emission* at microwave frequencies could oscillate in a resonant cavity and, in 1954, he built his first device. It will have thus taken almost forty years to move from Einstein's imagination to the practical realization of stimulated emission. Then, by replacing the incident radiation of light, of frequencies of 10^6 Ghz, by an incident radiation of microwaves of frequencies of the order of GHz, Charles Townes understood that the probability of spontaneous fallout of atoms would be divided by a million cubed. So, he chose to deflect the atoms being in the E_1 level against a wall cooled with liquid nitrogen, to solidify them. The atoms in the E_2 state were, on the other hand, directed towards an orifice to fall into a waveguide, in which was the incident wave. Subsequently, the emitted radar wave emerged amplified. Charles Townes called his device Microwave Amplification by Stimulated Emission of Radiation[7] (MASER).

In 1955, Charles Townes benefited from a sabbatical year during which he spent several months in a laboratory of the ENS on rue d'Ulm in Paris with Albert Kastler. In this laboratory, Kastler used optical pumping to move atoms from one energy level to another. Townes had the presentiment that it was possible to generalize optical pumping to build a LASER. According to Hecht (2010):

> "At this point Townes had essentially formulated a physics problem — how could one build an optical oscillator to generate coherent light by amplifying stimulated emission? Gould (then a 37-year-old doctoral student) had always dreamed of being an inventor, and had the advantage of having earlier worked with optics. He holed up in his apartment with a stack of references, coined the word *laser* (see Figure 7.3) for his invention, and sketched out a plan for the now-familiar Fabry–Pérot resonator in a notebook he had notarized on November 13, 1957. That notebook would become the foundation for a battle over patents, which after 30 years finally made Gould a multimillionaire. Townes teamed with Arthur Schawlow, a former Columbia colleague who had married Townes's sister and had worked on optical spectroscopy. Together they wrote a detailed proposal for what they called an "optical maser" that *Physical Review* published[8] in December 1958."

---

[7]See Gordon (1954).
[8]See Schawlow and Townes (1958).

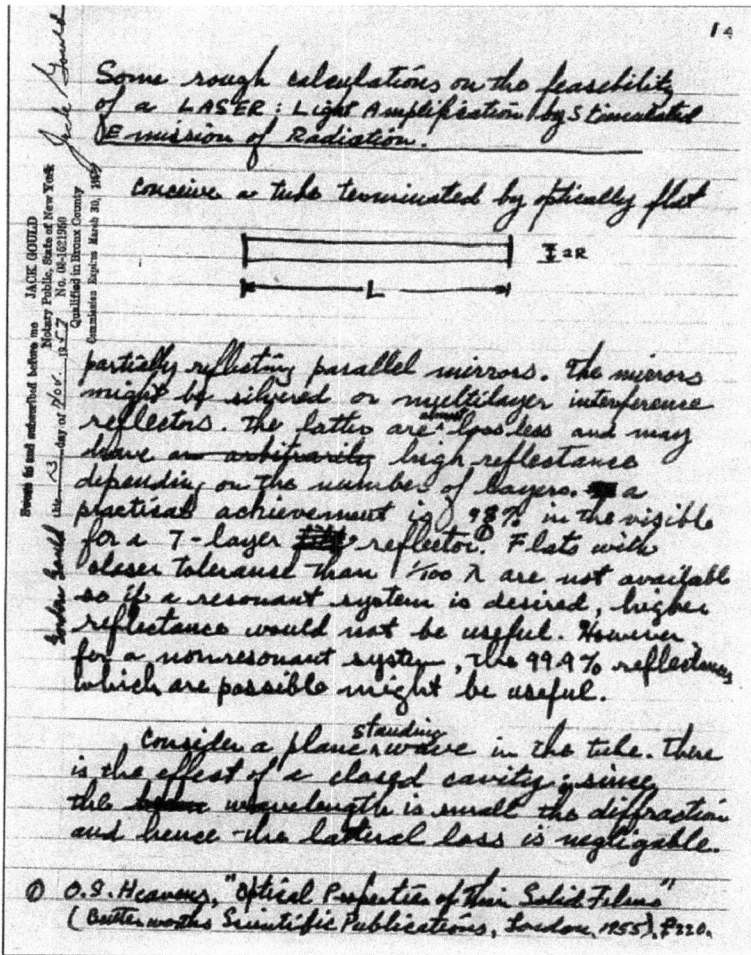

Figure 7.3. First page of the notebook wherein Gordon Gould coined the acronym LASER, and described the elements required to construct one.

The publication of the theoretical article by Townes and Schawlow then triggered a "laser race", in which many laboratories participated. In the beginning of 1960, Theodore Maiman, then working at Hughes Research laboratories, started investigating ruby doped with chromium ions. The pumping is triggered by a flash of white light. The yellow, green and purple radiation from the flash is absorbed by the chromium ions which change in energy level. A very large part of the white light spectrum is thus used to raise the atoms into higher electronic levels (the continuum). Unfortunately, as recalled by Hecht (2010):

> "Schawlow had decided ruby would not work in lasers because it was a three-level system, with its red line dropping to the ground state, and because other measurements had shown its red fluorescence was inefficient. Maiman made his own measurements and found that ruby fluorescence actually was quite efficient."

Maiman had the idea of making a pulse source and not a continuous one, for which the oscillation conditions are only realized in transient state. The ruby laser he developed was a surprisingly simple device to build and use. Due to the bandwidth of the doped crystal, only 10% of the white light is sufficient to carry out the population inversion. The atoms in the continuum move spontaneously, by energy exchange with the thermal vibrations in the crystal, towards the bottom of the energy band. This is how population inversion is achieved.

> "Maiman tested his design on May 16, 1960, by gradually increasing the voltage applied to the flashlamp until the pulses of red light grew sharply brighter and their time and spectral profiles showed the changes expected from a laser.
>
> Hughes chose to announce the laser at a July 7, 1960, press conference in New York after *Physical Review Letters* summarily rejected Maiman's report of the discovery.
>
> Maiman published a very short description of his experiment in *Nature*,[9] but the most complete account of his experiments did not appear[10] until 1961.[11]"

In December 1960, Ali Javan at Bell Labs built the first continuous-wave laser and the first gas-laser in which the discharge excited helium atoms, which transferred energy to the neon atoms that emitted light.[12]

---

[9] See Maiman (1960).
[10] See Maiman (1961) and Maiman *et al.* (1961).
[11] See Hecht (2010).
[12] See Javan *et al.* (1961).

7.4   LASER applications

As recalled by Hecht (2010):

> "Soon after Maiman built the first laser, his assistant Irnee
> D'Haenens joked that the laser was
>
> "a solution looking for a problem."
>
> Like any successful wisecrack, it contained a bit of truth. The
> laser was not a device invented to fill specific application re-
> quirements, like the telephone. It was more a discovery than
> an invention, a way to generate coherent light that laser devel-
> opers expected would find applications in broad areas, such as
> research or communications."

Thus, after its discovery, thanks to applied and industrial research, the
laser passed in a few decades from the status of a laboratory object to that of
a commercial "press-button" device with constantly improved performance.
Hecht (2010) explains that:

> "As new types of lasers and new applications emerged, the laser
> caught the public imagination. It had the good fortune to be
> invented when the public welcomed new technology with open
> arms and optimism. The United States was in the midst of
> a technology boom, and with the notable exception of nuclear
> weapons, the public generally saw new technology as bringing
> hope. A 1962 article titled "The Incredible Laser" gives a snap-
> shot of the laser's public image at the time. It promised "an
> exciting report on science's new 'Aladdin's lamp.' It can light
> up the moon, kill instantly, or perform miracle surgery"."

This is probably the reason why a laser seems to appear in the 1964
Goldfinger spy film, the third instalment in the James Bond series. As
recalled above, lasers did not exist in 1959 when the book of the same
name was written by Ian Fleming, nor did high-power industrial lasers at
the time the film was made, making them a novelty. In the novel, *Goldfinger*
uses a circular saw to try to kill Bond, but the film makers changed it to a
laser to make the film feel fresher.

Why a laser, specifically? Because in 1964, a laser was cutting edge (pun
intended) technology. Lasers were invented in 1960; when *Goldfinger* was
filmed, scientists still had not figured out a practical use for them. Giving
Goldfinger a laser for his deathtrap gave the scene a science-fiction flavor

and made it more exciting for audiences. The laser's inclusion also wound up giving birth to a whole new Bond cliché.

Bond is captured and strapped to a table with an overhead industrial laser, the beam slicing toward him (see Figure 7.4).

Figure 7.4.   James Bond (Sean Connery) in *Goldfinger* movie film.

To ensure that audiences would understand the laser, authors of the scene wrote *Goldfinger* some dialogue explaining the machine and how it works before activating it:

Auric Goldfinger explains:

> "You are looking at an industrial laser, which admits an extraordinary light not to be found in nature. It can project a spot on the moon, or at closer range, cut through solid metal. I will show you."

> James Bond: I think you made your point. Thank you for the demonstration.

> Auric Goldfinger: Choose your next witticism carefully Mr. Bond, it may be your last.

> James Bond: You expect me to talk?

> Auric Goldfinger: No Mr. Bond I expect you to die!

The laser scene was the first to be filmed for *Goldfinger* and one of the trickiest to complete. For starters, the laser itself was a dangerous tool not conducive for filming. Special effects supervisor Cliff Culley said:

"They bought a real laser in, which looked great. It had a pencil-thin line, but as soon as you turned all the studio lights on, it disappeared."

In his book, effects technician Albert Luxford (2002) added:

> "The laser was an extremely dangerous thing insofar as it having 400 to 500 volts going through the coil ... If you touched it, you'd have been dead. It wasn't a toy. If you'd gone within a foot of it when it was on, you'd have had arcs too - giving a very nasty shock, to say the least."

For these reasons, the real blue beam fired from the machine was not used; the orange laser beam in the film was added as an optical effect in post-production.

Unable to use the real laser to cut through the table, the crew improvised. During shooting, Luxford sat under the table and cut from underneath it with a blowtorch. In the scene, Bond mentioning Goldfinger's "Operation: Grand Slam" is what gets him out of a clean-cut death; Sean Connery saying these words was Luxford's cue to stop cutting. Luxford (2002) recalled:

> As I got nearer and nearer to his crotch, Sean was sweating a bit ... I was about three inches from his crotch when I stopped."

The laser scene is one of the most iconic in 007 history. As deathtraps come and go, Goldfinger's laser remains in the public memory.

7.5 References

Newton I. (1704). *Opticks or, a Treatise of the Reflexions, Refractions, Inflexions and Colours of Light. Also Two Treatises of the Species and Magnitude of Curvilinear Figures* (Samuel Smith and Benjamin Walford, London).

Maxwell J.C. (1862). On physical lines of force, *Philosophical Magazine*, Vol. XXIII, Ser. 4, pp. 12-24 & 85-95.

Einstein A. (1906). Theorie der Lichterzeugung und Lichtabsorption (On a Heuristic Point of View Concerning the Production and Transformation of Light), *Annalen der Physik*, 20(4), pp. 199-206.

Einstein A. (1916). Strahlungs -emission und -absorption nach der Quantentheorie (Emission and Absorption of Radiation in Quantum Theory), *Verhandlungen der Deutschen Physikalischen Gesellschaft*, 18, pp. 318-323.

Einstein A. (1917). Zur Quantentheorie der Strahlung (On the Quantum Theory of Radiation), *Physikalische Zeitschrift*, 18, pp. 121-128

Gordon J.P., Zeiger H.J. & Townes C.H. (1954). Molecular microwave oscillator and new hyperfine structure in the microwave spectrum of NH3, *Physical Review*, 95, pp. 282-284.

Hecht J. (2010). A short history of laser development, *Applied Optics*, 49, F99-F122.

Schawlow A.L. & Townes C.H. (1958). Infrared and optical masers, *Physical Review*, 112, pp. 1940-1949.

Maiman T.H. (1960). Stimulated optical radiation in ruby, *Nature*, 187, 493.

Maiman T.H. (1961). Stimulated optical emission in fluorescent solids, Part I. Theoretical considerations, *Physical Review*, 125, pp. 1145-1150.

Maiman T.H., Hoskins R.H., D'Haenens I.J., Asawa C.K. & Evtuhov V. (1961). Stimulated optical emission in fluorescent solids II, Spectroscopy and stimulated emission in ruby, *Physical Review*, 125, pp. 1151-1157.

Javan A., Bennett Jr. W.R. & Herriott D.R. (1961). Population inversion and continuous optical maser oscillation in a gas discharge containing a He-Ne mixture, *Physical Review Letters*, 6(3), pp. 106-110.

Luxford A. (2002). *Gimmick Man: Memoir of a Special Effects Maestro* (McFarland & Company); Illustrated edition.

Appendix A

Appendix

In this appendix we present three different kinds of normalization of the basic laser equations. In the first section, we briefly recall the laser rate equations and its stationary solutions.

A.1 Laser rate equations

The laser rate equations reads:

$$\dot{n} = -Kn + Gn\Delta$$
$$\dot{\Delta} = -\gamma_{\parallel} \left(\Delta - \Delta_0 \right) - 2Gn\Delta \tag{A.1}$$

where

- n represents the photon number,

- Δ is the population inversion, i.e. $N_2 - N$,

- K is the decay rate of the laser intensity,

- $\gamma_{\parallel}$ is the decay rate of the population inversion,

- Δ_0 is the population inversion imposed by the pump,

- G is the field-matted coupling constant (stimulated emission).

A.1.1 *Stationary solutions*

The stationary solutions of the above equations are obviously given by the nullclines process:

$$\dot{n} = 0$$
$$\dot{\Delta} = 0 \tag{A.2}$$

The first equation of (A.2) gives:

$$n\left(-K + G\Delta\right) = 0 \quad \rightarrow \quad \Delta_s = \frac{K}{G} = \Delta_{threshold} = \Delta_{th} \tag{A.3}$$

The second equation of (A.2) leads to:

$$2Gn_s\Delta_s = -\gamma_{\|}\left(\Delta - \Delta_0\right) \quad \rightarrow \quad n_s = -\frac{\gamma_{\|}}{2G}\frac{\Delta - \Delta_0}{\Delta_s} \tag{A.4}$$

Thus, we have:

$$n_s = \frac{\gamma_{\|}}{2G}\left(\frac{\Delta_0}{\Delta_s} - 1\right) \tag{A.5}$$

By posing $I_{sat} = \dfrac{\gamma_{\|}}{2G}$ we obtain:

$$n_s = I_{sat}\left(\frac{\Delta_0}{\Delta_s} - 1\right) = I_{sat}\left(A - 1\right) \tag{A.6}$$

where $A = \dfrac{\Delta_0}{\Delta_{th}}$ represents the pump parameter (normalized threshold).

A.2 First normalization

Let us take:

$$X = \frac{n}{I_{sat}} = \frac{n}{\gamma_{\|}/2G}$$
$$Y = \frac{\Delta}{\Delta_{th}} = \frac{\Delta}{K/G} \tag{A.7}$$

we obtain:

$$\dot{X} = -KX\left(1 - Y\right)$$
$$\dot{Y} = -\gamma_{\|}\left(Y - A + XY\right) \tag{A.8}$$

A.2.1 *Stationary solutions*

The stationary solutions of the above equations are obviously given by the nullclines process:

$$\dot{X} = 0$$
$$\dot{Y} = 0 \tag{A.9}$$

These equations give first:

$$\bar{X} = A - 1$$
$$\bar{Y} = 1 \tag{A.10}$$

where $\bar{X}$ is here positive defined. Then, we also have:

$$\bar{X} = 0$$
$$\bar{Y} = A \tag{A.11}$$

where $0 < A < 1$ (0 int. solution). It follows from Eqs. (A.11) that:

$$\bar{X} \neq 0 = A - 1$$
$$\bar{Y} = 1 \tag{A.12}$$

where $A > 1$ is a non-zero solution.

A.2.2 *Linear stability analysis*

Let us evaluate the Jacobian matrix of system (A.8) at the non-zero solution $(\bar{X} \neq 0 = A - 1, \bar{Y} = 1)$, we have:

$$J = \begin{pmatrix} -K + K\bar{Y} & K\bar{X} \\ -\gamma_{\|}\bar{Y} & -\gamma_{\|} - \gamma_{\|}\bar{X} \end{pmatrix} = \begin{pmatrix} 0 & K(A-1) \\ -\gamma_{\|} & -\gamma_{\|}A \end{pmatrix} \tag{A.13}$$

The characteristic polynomial reads then:

$$\lambda^2 + \gamma_{\|}A\lambda + K\gamma_{\|}(A-1) = 0 \tag{A.14}$$

The roots of this equation are:

$$\lambda_{1,2} = -\frac{\gamma_\| A}{2} \pm \frac{\sqrt{(\gamma_\| A)^2 - 4K\gamma_\| (A-1)}}{2},$$

$$\lambda_{1,2} = -\frac{\gamma_\| A}{2} \pm i\frac{\sqrt{4K\gamma_\| (A-1) - (\gamma_\| A)^2}}{2}$$

(A.15)

Since $A > 1$, $\bar{X}$ is a stable solution.

Now, let us evaluate the Jacobian matrix of system (A.13) at the non zero solution $(\bar{X} = 0, \bar{Y} = A)$, we have:

$$J = \begin{pmatrix} -K + K\bar{Y} & K\bar{X} \\ -\gamma_\| \bar{Y} & -\gamma_\| - \gamma_\| \bar{X} \end{pmatrix} = \begin{pmatrix} -K + KA & 0 \\ -\gamma_\| A & -\gamma_\| \end{pmatrix}$$

(A.16)

The characteristic polynomial reads then:

$$\lambda^2 + (K - KA + \gamma_\|)\lambda + \gamma_\| (K - KA) = 0$$

(A.17)

The roots of this equation are:

$$\lambda_{1,2} = -\frac{K + KA + \gamma_\|}{2} \pm \frac{\sqrt{(K(1-A) + \gamma_\|)^2 - 4K\gamma_\| (1-A)}}{2}$$

(A.18)

If $A < 1$, $\lambda_{1,2}$ are both negative and the solution is stable.

If $A = 1$, $\lambda_1 = 0$ and $\lambda_2 = -\gamma_\|$. This solution corresponds to the marginal stability.

If $A > 1$, $\lambda_1 > 0$ and $\lambda_2 < 0$, the solution is unstable.

A.3 Second normalization

Let us take in (A.1):

$$X = \frac{n}{K/G}$$

$$Y = \frac{\Delta}{K/G}$$

(A.19)

we obtain:

$$\dot{X} = -KX\,(1 - Y)$$
$$\dot{Y} = -\gamma_{||}\,(Y - A) - 2KXY$$

$$(A.20)$$

A.3.1 *Stationary solutions*

As previously, by taking $\dot{X} = 0$ and $\dot{Y} = 0$, we find:

$$\bar{X} = 0$$
$$\bar{Y} = A$$

$$(A.21)$$

A.3.2 *Linear stability analysis*

Let us evaluate the Jacobian matrix of system (A.20) at the non-zero solution $(\bar{X} = 0, \bar{Y} = A)$, we have:

$$J = \begin{pmatrix} -K + K\bar{Y} & K\bar{X} \\ -2K\bar{Y} & -\gamma_{||} - 2\gamma_{||}\bar{X} \end{pmatrix} = \begin{pmatrix} -K + KA & 0 \\ -2KA & -\gamma_{||} \end{pmatrix} \qquad (A.22)$$

The characteristic polynomial reads then:

$$\lambda^2 + \left(K - KA + \gamma_{||}\right)\lambda + K\gamma_{||}\,(1 - A) = 0 \qquad (A.23)$$

The roots of this equation are:

$$\lambda_{1,2} = -\frac{K - KA + \gamma_{||}}{2} \pm \frac{\sqrt{\left(K - KA + \gamma_{||}\right)^2 - 4K\gamma_{||}\,(1 - A)}}{2} \qquad (A.24)$$

If $0 < A < 1$, both $\lambda_{1,2}$ are negative and so, the solution is a stable.

If $A = 1$, $\lambda_1 = 0$ and $\lambda_2 = -\gamma_{||}$. This solution corresponds to the marginal stability.

If $A > 1$, $\lambda_1 > 0$ and $\lambda_2 < 0$, the solution is unstable.

Now, let us evaluate the Jacobian matrix of system (A.13) at the non-zero solution $(\bar{X} = \gamma_{\parallel}/2K(A-1), \bar{Y} = 1)$, we have:

$$J = \begin{pmatrix} -K + K\bar{Y} & K\bar{X} \\ -2K\bar{Y} & -\gamma_{\parallel} - 2K\bar{X} \end{pmatrix} = \begin{pmatrix} 0 & \gamma_{\parallel}/2\,(A-1) \\ -2K & -\gamma_{\parallel} - \gamma_{\parallel}\,(A-1) \end{pmatrix} \quad (A.25)$$

The characteristic polynomial reads then:

$$\lambda^2 + \left(\gamma_{\parallel} + \gamma_{\parallel}\,(A-1)\right)\lambda + \gamma_{\parallel}K\,(A-1) = 0 \quad (A.26)$$

The roots of this equation are:

$$\lambda_{1,2} = -\frac{\gamma_{\parallel}A}{2} \pm i\frac{\sqrt{4K\gamma_{\parallel}\,(A-1) - \left(\gamma_{\parallel}A\right)^2}}{2} \quad (A.27)$$

Both $\lambda_{1,2}$ are negative and so, the solution is a stable.

A.4   Third normalization

Let us take $X' = \alpha X$ in (A.20), we obtain:

$$\begin{aligned} \dot{X}' &= -KX'\,(1-Y) \\ \dot{Y} &= -\gamma_{\parallel}\,(Y-A) - 2\frac{K}{\alpha}X'Y \end{aligned} \quad (A.28)$$

Obviously, if $\alpha = 2K/\gamma_{\parallel}$, we find again the first normalization.

A.4.1   *Stationary solutions*

By taking $\dot{X}' = 0$ and $\dot{Y} = 0$, we find:

$$\begin{aligned} \bar{X} &= \frac{\alpha\gamma_{\parallel}}{2K}\,(A-1) \\ \bar{Y} &= 1 \end{aligned} \quad (A.29)$$

or,

$$\begin{aligned} \bar{X} &= 0 \\ \bar{Y} &= A \end{aligned} \quad (A.30)$$

A.4.2 *Linear stability analysis*

The Jacobian matrix of system (A.28) reads:

$$J = \begin{pmatrix} -K + K\bar{Y} & K\bar{X} \\ \dfrac{2K}{\alpha}\bar{Y} & -\gamma_{\|} - \dfrac{2K}{\alpha}\bar{X} \end{pmatrix} \tag{A.31}$$

Let us first evaluate the Jacobian matrix for the fixed point (A.29), we obtain

$$J = \begin{pmatrix} 0 & \dfrac{\alpha\gamma_{\|}}{2}\left(A - 1\right) \\ -\dfrac{2K}{\alpha} & -\gamma_{\|} - \gamma_{\|}\left(A - 1\right) \end{pmatrix} \tag{A.32}$$

The characteristic polynomial reads then:

$$\lambda^2 + \gamma_{\|}A\lambda + \gamma_{\|}K\left(A - 1\right) = 0 \tag{A.33}$$

The roots of this equation are:

$$\lambda_{1,2} = -\frac{\gamma_{\|}A}{2} \pm i\frac{\sqrt{4K\gamma_{\|}\left(A - 1\right) - \left(\gamma_{\|}A\right)^2}}{2} \tag{A.34}$$

Both real parts of $\lambda_{1,2}$ are negative and so, the solution is a stable.

Now, let us first evaluate the Jacobian matrix for the fixed point (A.30), we obtain

$$J = \begin{pmatrix} -K + KA & 0 \\ -\dfrac{2KA}{\alpha} & -\gamma_{\|} \end{pmatrix} \tag{A.35}$$

The characteristic polynomial reads then:

$$\lambda^2 + \left(K\left(A - 1\right) + \gamma_{\|}\right)\lambda + \gamma_{\|}K\left(1 - A\right) = 0 \tag{A.36}$$

The roots of this equation are:

$$\lambda_{1,2} = -\frac{\left(K\left(A - 1\right) + \gamma_{\|}\right)}{2} \pm \frac{\sqrt{\left(K\left(A - 1\right) + \gamma_{\|}\right)^2 - 4K\gamma_{\|}\left(1 - A\right)}}{2} \tag{A.37}$$

Both roots $\lambda_{1,2}$ are negative and so, the solution is stable.

Appendix B

Appendix

This appendix presents a method allowing to provide a upper bound for the Hopf bifurcation parameter for any three-dimensional autonomous dynamical system for which the fixed point coordinates cannot be easily expressed analytically as it is the case for the dynamical system (3.2) for which the coordinates of the fixed point I_2 are the roots of a cubic polynomial. Let us suppose that the three eigenvalues of the Jacobian matrix J of this dynamical system evaluated at the fixed point (I_2 in our case) are real and complex conjugate λ_1, $\lambda_{2,3} = \alpha \pm i\omega$. The Cayley–Hamilton eigenpolynomial reads:

$$\lambda^3 - \sigma_1\lambda^2 + \sigma_2\lambda - \sigma_3 = 0 \tag{B.1}$$

where $\sigma_1 = Tr\,(J)$, $\sigma_2 = \sum_{i=1}^{3} M_{ii}\,(J)$ is the sum of all first-order diagonal minors of J and $\sigma_3 = Det\,(J)$. Thus, we have:

$$\sigma_1 = Tr\,(J) = \lambda_1 + \lambda_2 + \lambda_3 = \lambda_1 + 2\alpha,$$

$$\sigma_2 = \sum_{i=1}^{3} M_{ii}\,(J) = \lambda_1\lambda_2 + \lambda_1\lambda_3 + \lambda_2\lambda_3 = 2\alpha\lambda_1 + \beta, \tag{B.2}$$

$$\sigma_3 = Det\,(J) = \lambda_1\lambda_2\lambda_3 = \lambda_1\beta,$$

where $\beta = \alpha^2 + \omega^2$. In order to analyze the stability of fixed points according to a control parameter value (B_0 here), i.e. the occurrence of Hopf bifurcation, we propose to use the Routh–Hurwitz' theorem [Routh (1877); Hurwitz (1893)] which states that if $D_1 = \sigma_2$ and $D_2 = \sigma_3 - \sigma_2\sigma_1$ are both positive then *eigenpolynomial* equation (B.1) would have eigenvalues with negative real parts. From Eqs. (B.2) it can be stated that:

$$\alpha = \frac{\sigma_1\sigma_2 - \sigma_3}{\lambda_1^2 + \sigma_2} \tag{B.3}$$

Thus, $\alpha = 0$ provided that $\kappa = \sigma_1\sigma_2 - \sigma_3 = 0$. For dynamical system (3.2), we obtain:

$$\kappa = (1 + x)\left[p_0 xy\epsilon_1 + (1 + x + \epsilon_2)\epsilon_2\right] - 2B_1 k_1 xz\epsilon_1\epsilon_2^2 \qquad \text{(B.4)}$$

then, by replacing x, y and z the coordinates (3.3) of the fixed point I_2, we have:

$$B_0 = \frac{1 + \epsilon_2}{2B_1 k_1 x^* \epsilon_1 \epsilon_2} + \frac{p_0\epsilon_1 + \epsilon_2(2 + \epsilon_2)}{2B_1 k_1 \epsilon_1 \epsilon_2^2} + \frac{(1 + 2B_1^2 k_1 \epsilon_1 \epsilon_2)x^*}{2B_1 k_1 \epsilon_1 \epsilon_2} \qquad \text{(B.5)}$$

Positivity of fixed points has led to $x^* \leqslant p_0 - 1$ which implies that $\max(x^*) = p_0 - 1$. Thus, by taking $x^* = p_0 - 1$ in (B.5) and while using the parameters sets of our experiment, i.e. $\epsilon_1 = 200$, $\epsilon_2 = 6$, $k_1 = 12$, $p_0 = 1.208$ and $B_1 = 0.555$, we find:

$$B_0^{Hopf} \leqslant 0.12057$$

The numerical computation of the Hopf bifurcation parameter value has been found equal to 0.12036 which is below and very near the upper bound analytically obtained.

Bibliography

Abraham N.B., Narducci L.M. & Mandel P. (1988). *Progress in Optics, Volume XXV* (ed. by H. Haken, Elsevier). (See also articles in "Selected papers on optical chaos," SPIE Milestone Series Vol. MS75, eds. F.T. Arecchi and R.O. Harrison, SPIE Optical Engineering Press (1993)).

Aguirre J., d'Ovidio F. & Sanjuán M.A.F. (2004). Controlling chaotic transients: Yorke's game of survival, *Physical Review E*, 69, 016203.

Al-Naimee K., Marino F., Ciszak M., Meucci R. & Arecchi F.T. (2009). Chaotic spiking and incomplete homoclinic scenarios in semiconductor lasers with optoelectronic feedback, *New J. Phys.*, 11, 073022.

Andronov A.A. & Chaikin S.E. (1937). *Theory of Oscillators*, Moscow, I., English Translation; (Princeton University Press, Princeton, NJ, USA), 1949.

Andronov A.A., Leontovich E., Gordon I. & Maier A. (1971). *Theory of Bifurcations of Dynamical Systems on a Plane* (Israel Program for Scientific Translations, Jerusalem).

D'Alembert J. (1748). Suite des recherches sur le calcul intégral, quatrième partie : méthodes pour intégrer quelques équations différentielles, *Hist. Acad. Berlin*, tome IV, 275-291.

Arecchi F.T. (1965). Measurement of the statistical distribution of Gaussian and laser sources, *Physical Review Letters*, 15(24), pp. 912-916.

Arecchi F.T. & Bonifacio R. (1965). Theory of optical maser amplifiers, *IEEE Journal of Quantum Electronic*, 1(4), pp. 169-178.

Arecchi F.T., Meucci R., Puccioni G.P. & Tredicce J.R. (1982). Experimental evidence of subharmonic bifurcations, multistability, and turbulence in a Q-switched gas laser, *Physical Review Letters*, 49(17), pp. 1217-1220.

Arecchi F.T., Lippi G.L., Puccioni G. & Tredicce J.R. (1984). Deterministic chaos in laser with injected signal, *Opt. Commun.*, 51(5), pp. 308-314.

Arecchi F.T., Gadomski W. & Meucci R. (1986). Generation of chaotic dynamics by feedback on a laser, *Phys. Rev. A*, 34(2), pp. 1617-1620.

Arecchi F.T., Meucci R. & Gadomski W. (1987). Laser dynamics with competing instabilities, *Phys. Rev. Lett.*, 58(21), pp. 2205-2208.

Arecchi F.T. (1987). *Instabilities and Chaos in Quantum Optics*, eds. F.T. Arecchi and R.O. Harrison (Springer, Berlin).

Arecchi F.T., Fortuna L., Frasca M., Meucci R. & Sciuto G. (2005). A programmable electronic circuit for modelling CO_2 laser dynamics, *Chaos*, 15, 043104.

Arecchi F.T. & Kurths J. (2009). Introduction to focus issue: Nonlinear dynamics in cognitive and neural systems, *Chaos*, 19, 015101.

Bender C.M. & Orszag S.A. (1999). *Advanced Mathematical Methods for Scientists and Engineers* (Springer, New York, NY, USA).

Brøns M. & Bar-Eli K. (1994). Asymptotic analysis of canards in the EOE equations and the role of the inflection line, In *Proceedings of the Royal Society of London, Series A: Mathematical, Physical and Engineering Sciences*, Vol. 445, pp. 305-322.

Boccaletti S., Grebogi C., Lai Y.C., Mancini H. & Maza D. (2000). Control of chaos: Theory and applications, *Phys. Rep.*, 329(3), pp. 103-197.

Buscarino A., Fortuna L. & Frasca M. (2014). The jerk dynamics of Chua's circuit, *Int. J. Bifurcation Chaos*, 24(06), 1450085.

Celikovshy S. & Chen G. (2002). On a generalized Lorenz canonical form of chaotic systems, *Int. J. Bifurcation Chaos*, 12, pp. 1789-1812.

Chen G. (2020). Generalized Lorenz systems family, arXiv:2006.04066.

Chua L.O., Kumaro M. & Matsumoto T. (1986). The double scroll family, *IEEE Transactions on Circuits and Systems*, 33, pp. 1072-1118.

Ciofini M., Politi A. & Meucci R. (1993). Effective two-dimensional model for CO_2 lasers, *Phys. Rev. A*, 48, pp. 605-610.

Cole J.D. (1968). *Perturbation Methods in Applied Mathematics* (Blaisdell, Waltham, MA, USA).

Darboux G. (1878). Mémoire sur les équations différentielles algébriques du premier ordre et du premier degré, *Bull. Sci. Math.*, Sér. 2, p. 60-96, p. 123-143 & p. 151-200.

Donati S. & Mirasso C.R. (2002). Feature section on optical chaos and applications to cryptography, *IEEE J. Quantum Electron.*, 38, pp. 1138-1205.

Eckmann J.P. & Ruelle D. (1985). Ergodic theory of chaos and strange attractors, *Rev. Mod. Phys.*, 57, pp. 617-656.

Einstein A. (1906). Theorie der Lichterzeugung und Lichtabsorption (On a Heuristic Point of View Concerning the Production and Transformation of Light), *Annalen der Physik*, 20(4), pp. 199-206.

Einstein A. (1916). Strahlungs -emission und -absorption nach der Quantentheorie (Emission and Absorption of Radiation in Quantum Theory), *Verhandlungen der Deutschen Physikalischen Gesellschaft*, 18, pp. 318-323.

Einstein A. (1917). Zur Quantentheorie der Strahlung (On the Quantum Theory of Radiation), *Physikalische Zeitschrift*, 18, pp. 121-128.

Fenichel N. (1971) Persistence and smoothness of invariant manifolds for flows, *Ind. Univ. Math. J.*, 21, pp. 193-225.

Fenichel N. (1974). Asymptotic stability with rate conditions, *Ind. Univ. Math. J.*, 23, pp. 1109-1137.

Fenichel N. (1977). Asymptotic stability with rate conditions II, *Ind. Univ. Math. J.*, 26, pp. 81-93.

Fenichel N. (1979). Geometric singular perturbation theory for ordinary differential equations, *J. Diff. Eq.*, 31, pp. 53-98.

Feudel U., Grebogi C., Hunt B.R. & Yorke J.A. (1996). Map with more than 100 coexisting low-period periodic attractors, *Phys. Rev. E*, 54, pp. 71-81.

Feudel U. (2008). Complex dynamics in multistable systems, *Int. J. Bifurcation Chaos*, 18(6), pp. 1607-1626.

Feudel U., Pisarchik A.N. & Showalter K. (2018). Multistability and tipping: From Mathematics and physics to climate and brain — Minireview and preface to the focus issue, *Chaos*, 28, 033501 (8 pages).

Fischer I., Liu Y. & Davis P. (2000). Synchronization of chaotic semiconductor laser dynamics on subnanosecond time scales and its potential for chaos communication, *Phys. Rev. A*, 62(1), 011801 (4 pages).

Freire J.G., Meucci R., Arecchi F.T. & Gallas J.A.C. (2015). Self-organization of pulsing and bursting in a CO_2 laser with opto-electronic feedback, *Chaos*, 25(9), 097607.

Gear C.W., Kaper T.J., Kevrekidis I.G. & Zagaris A. (2005). Projecting to a slow manifold: Singularly perturbed systems and legacy codes, *SIAM J. Appl. Dyn. Syst. Math.*, 4, pp. 711-732.

Ginoux J.M. & Rossetto B. (2006). Slow manifold of a neuronal bursting model, In *Emergent Properties in Natural and Articial Dynamical Systems*, eds. M.A. Aziz-Alaoui and C. Bertelle (Springer, Berlin/Heidelberg, Germany), pp. 119-128.

Ginoux J.M. & Rossetto B. (2006). Differential geometry and mechanics applications to chaotic dynamical systems, *Int. J. Bifurcation Chaos*, 4, pp. 887-910.

Ginoux J.M., Rossetto B. & Chua L.O. (2008). Slow invariant manifolds as curvature of the flow of dynamical systems, *Int. J. Bifurcation Chaos*, 11, pp. 3409-3430.

Ginoux J.M. (2009). *Differential Geometry Applied to Dynamical Systems* World Scientific Series on Nonlinear Science, Series A, Vol. 66 (World Scientific, Singapore).

Ginoux J.M. & Llibre J. (2011). The flow curvature method applied to canard explosion, *J. Phys. A Math. Theor.*, 44, 465203.

Ginoux J.M., Llibre J. & Chua L.O. (2013). Canards from Chua's circuit, *Int. J. Bifurcation Chaos*, 23, 1330010.

Ginoux J.M. (2014). The slow invariant manifold of the Lorenz–Krishnamurthy model, *Qual. Theory Dyn. Syst.*, 13, pp. 19-37.

Ginoux J.M. & Llibre J. (2015). Canards existence in FitzHugh–Nagumo and Hodgkin–Huxley neuronal models, *Math. Probl. Eng.*, 2015, 342010.

Ginoux J.M. & Llibre J. (2016). Canards existence in memristor's circuits, *Qual. Theory Dyn. Syst.*, 15, pp. 383-431.

Ginoux J.M., Llibre J. & Tchizawa K. (2019). Canards existence in the Hindmarsh–Rose model, *Math. Model. Nat. Phenom.*, 14, pp. 1-21.

Ginoux J.M. (2021). Slow invariant manifolds of slow-fast dynamical systems, *Int. J. Bifurcation Chaos*, 31, 2150112-1-17.

Ginoux J.M., Lebiedz D., Meucci R. & Llibre J. (2022). Flow curvature manifold and energy of generalized Liénard systems, *Chaos, Solitons & Fractals*, 161, 112354.

Gordon J.P., Zeiger H.J. & Townes C.H. (1954). Molecular microwave oscillator and new hyperfine structure in the microwave spectrum of NH3, *Physical Review*, 95, pp. 282-284.

Goswami B.K. & Pisarchick A.N. (2008). Controlling multistability by small periodic perturbation, *Int. J. Bifurcation Chaos*, 18(6), pp. 1645-1673.

Grebogi C., Ott E. & Yorke, J. (1982). Chaotic attractors in crisis, *Phys. Rev. Lett.*, 48(22), pp. 1507-1510.

Grebogi C., Ott E. & Yorke J.A. (1983). Crises: Sudden changes in chaotic attractors and chaotic transients, *Physica D*, 7, pp. 181-200.

Grebogi C., Ott E., Romeiras F. & Yorke J. (1987). Critical exponents for crisis-induced intermittency, *Phys. Rev. A*, 36(11), pp. 5365-5380.

Haag M. (1879). Note sur les relations entre les éléments caractéristiques d'une courbe gauche et les accélérations du point qui la décrit, *Bulletin de la S. M. F.*, tome 7, pp. 140-143.

Haken H. (1975). Analogy between higher instabilities in fluids and lasers, *Physics Letters A*, 53(1), pp. 77-78.

Haken H. (1985). *Light Volume 2: Laser Light Dynamics* (North Holland).

Hau Z., Kang N., Kong X., Chen G. & Yan G. (2010). On the equivalence of Lorenz system and Chen system, *Int. J. Bifurcation Chaos*, 20, pp. 557-560.

Hecht J. (2010). A short history of laser development, *Applied Optics*, 49, F99-F122.

Hirsch M.W., Pugh C.C. & Shub M. (1977). *Invariant Manifolds* (Springer, New York, NY, USA).

Hopf E. (1942). Abzweigung einer periodischen Lösung von einer stationären Lösung eines Differentialsystems, *Berichte der MathematischPhysikalischen Klasse der Sächsischen Akademie der Wissenschaften zu Leipzig*, Band XCIV, Sitzung vom 19. January 1942, pp. 3-22. See L.N. Howard and N. Kopell, A Translation of Hopf's Original Paper, pp. 163-193 and Editorial Comments, pp. 194-205 in J. Marsden and M. McCracken.

Hurwitz A. (1893). Über die Bedingungen, unter welchen eine Gleichung nur-Wurzeln mit negativen reellen Theilen besitzt, *Math. Ann.*, 41, pp. 403-442.

Jafari A., Hussain I., Nazarimehr F., Golpayegani S.M.R.H. & Jafari S. (2021). A simple guide for plotting a proper bifurcation diagram, *Int. J. Bifurcation*

Chaos, 31(1), 2150011 (11 pages).

Javan A., Bennett Jr. W.R. & Herriott D.R. (1961). Population inversion and continuous optical maser oscillation in a gas discharge containing a He-Ne mixture, *Physical Review Letters*, 6(3), pp. 106-110.

Klein M. & Baier G. (1991). Hierarchies of dynamical systems, In *A Chaotic Hierarchy*, eds. G. Baier and M. Klein (World Scientific, Singapore).

Kuznetsov Yu. A. (2004). *Elements of Applied Bifurcation Theory* (Springer-Verlag, New York), Third edition.

Lang R. & Kobayashi K. (1980). External optical feedback effects on semiconductor injection laser properties, *IEEE Journal of Quantum Electronics*, 16(3), pp. 347-355.

Levinson N. (1949). A second-order differential equation with singular solutions, *Ann. Math.*, 50, pp. 127-153.

Lima R. & Pettini M. (1990). Suppression of chaos by resonant parametric perturbation, *Phys. Rev. A*, 41(2), pp. 726-733.

Lorenz E.N. (1963). Deterministic non-periodic flows, *J. Atmos. Sci.*, 20, pp. 130-141.

Lotka A.J. (1910). Contribution to the theory of periodic reaction, *J. Phys. Chem.*, 14(3), pp. 271-274.

Lotka A.J. (1920). Analytical note on certain rhythmic relations in organic systems, *Proc. Natl. Acad. Sci. U.S.A.*, 6(7) pp. 410-415.

Lucarini V. & Bodai T. (2017). Edge states in the climate system: Exploring global instabilities and critical transitions, *Nonlinearity*, 30, R32-R66.

Luxford A. (2002). *Gimmick Man: Memoir of a Special Effects Maestro* (McFarland & Company); Illustrated edition.

Maiman T.H. (1960). Stimulated optical radiation in ruby, *Nature*, 187, 493.

Maiman T.H. (1961). Stimulated optical emission in fluorescent solids, Part I. Theoretical considerations, *Physical Review*, 125, pp. 1145-1150.

Maiman T.H., Hoskins R.H., D'Haenens I.J., Asawa C.K. & Evtuhov V. (1961). Stimulated optical emission in fluorescent solids II, Spectroscopy and stimulated emission in ruby, *Physical Review*, 125, pp. 1151-1157.

Marino I.P. & Míguez J. (2006). An approximate gradient-descent method for joint parameter estimation and synchronization of coupled chaotic systems, *Physics Letters A*, 351(4-5), pp. 262-267.

Marsden J. & McCracken M. (1976). *Hopf Bifurcation and its Applications* (Springer-Verlag, New York).

Maas U. & Pope S.B. (1992). Simplifying chemical kinetics: Intrinsic low-dimensional manifolds in composition space, *Combust. Flame*, 88, pp. 239-264.

Matsumoto T. (1984). A chaotic attractor from Chua's circuit, *IEEE Transactions on Circuits and Systems*, 31, pp. 1055-1058.

Maxwell J.C. (1862). On physical lines of force, *Philosophical Magazine*, Vol. XXIII, Ser. 4, pp. 12-24 & 85-95.

May R. (1977). Thresholds and breakpoints in ecosystems with multiplicity of stable states, *Nature*, 269, pp. 471-477.

McNeil B. (2015). Due credit for Maxwell-Bloch equations, *Nature Photonics*, 9, 207.

Meucci R., Poggi A., Arecchi F.T. & Tredicce J.R. (1988). Dissipativity of an optical chaotic system characterized via generalized multistability, *Opt. Commun.*, 65, pp. 151-156.

Meucci R., Gadomski W., Ciofini M. & Arecchi F.T. (1994). Experimental control of chaos by means of weak parametric perturbations, *Physical Review E*, 49(4), R2528-R2531.

Meucci R., Cinotti D., Allaria E., Billings L., Triandaf I., Morgan D. & Schwartz I.B. (2004). Global manifold control in a driven laser: Sustaining chaos and regular dynamics, *Physica D*, 189(1-2), pp. 70-80.

Meucci R., Allaria E., Salvadori F. & Arecchi F.T. (2005). Attractor selection in chaotic dynamics, *Physical Review Letters*, 95(18), 184101.

Meucci R., Euzzor S., Pugliese E., Zambrano S., Gallas M.R. & Gallas J.A.C. (2016). Optimal phase-control strategy for damped-driven Duffing oscillators, *Physical Review Letters* 116(4), 044101 (5 pages).

Meucci R., Euzzor S., Arecchi F.T. & Ginoux J.-M. (2021). Minimal universal model for chaos in a laser with feedback, *Int. J. Bifurcation Chaos*, 31(4), 2130013 (10 pages).

Meucci R., Ginoux J.M., Mehrabbeik M., Jafari S. & Sprott J.C. (2022). Generalized multistability and its control in a laser, *Chaos*, 32(8), 083111.

Meucci R. & Kurths J. (2022). In memoriam — Tito Arecchi (11 December 1933–15 February 2021), *Chaos*, 32, 080401.

Newton I. (1704). *Opticks or, a Treatise of the Reflexions, Refractions, Inflexions and Colours of Light. Also Two Treatises of the Species and Magnitude of Curvilinear Figures* (Samuel Smith and Benjamin Walford, London).

Ohtsubo J. & Davis P. (2005). Chaotic optical communication, In *Unlocking Dynamical Diversity-optical Feedback Effects on Semiconductor Lasers*, Chap. 10, eds. D. Kane and K.A. Shore (Wiley, Chichester).

O'Malley R.E. (1974). *Introduction to Singular Perturbations* (Academic Press, New York, NY, USA).

O'Malley R.E. (1991). *Singular Perturbations Methods for Ordinary Differential Equations* (Springer, New York, NY, USA).

Ott E. (1993). *Chaos in Dynamical Systems* (Cambridge University Press).

Ott E., Grebogi C. & Yorke J. (1990). Control chaos, *Phys. Rev. Lett.*, 64(11), pp. 1196-1199.

Pecora L.M. & Carroll T.L. (1990). Synchronization in chaotic systems, *Phys. Rev. Lett.*, 64, pp. 821-824.

Pisarchik A.N., Meucci R. & Arecchi F.T. (2000). Discrete homoclinic orbits in a laser with feedback, *Physical Review E*, 62(6), 8823.

Pisarchik A.N., Jaimes-Reategui R., Sevilla-Escoboza R., Huerta-Cuellar G. &

Taki M. (2011). Rogue waves in a multistable system, *Phys. Rev. Lett.*, 107, 274101 (5 pages).

Poincaré H. (1892, 1893, 1899). *Les méthodes Nouvelles de la Mécanique Céleste* (Gauthier-Villars, Paris, France), Volumes I, II & III.

Qu Z., Hu G., Yang G. & Qin G. (1995). Phase effect in taming nonautonomous chaos by weak harmonic perturbations, *Phys. Rev. Lett.*, 74, pp. 1736-1739.

Ricci L., Perinelli A., Castelluzzo M., Euzzor S. & Meucci R. (2021). Experimental evidence of chaos generated by a minimal universal oscillator model, *Int. J. Bifurcation Chaos*, 31(12), 2150205 (13 pages).

Roessler O.E. (1976). An equation for chaos, *Phys. Lett. A*, 57(5), pp. 397-398.

Rossetto B. (1986). Trajectoires lentes des systèmes dynamiques lents-rapides, In *Analysis and Optimization of System* (Springer, Berlin/Heidelberg, Germany), pp. 680-695.

Rossetto B. (1987). Singular approximation of chaotic slow-fast dynamical systems, In *The Physics of Phase Space Nonlinear Dynamics and Chaos Geometric Quantization, and Wigner Function* (Springer, Berlin/Heidelberg, Germany), pp. 12-14.

Rossetto B., Lenzini T., Ramdani S. & Suchey G. (1998). Slow fast autonomous dynamical systems, *Int. J. Bifurcation Chaos*, 8, pp. 2135-2145.

Routh E.J. (1877). *A Treatise on the Stability of a Given State of Motion: Particularly Steady Motion* (Macmillan and Co.).

Ryashko L. (2018). Sensitivity analysis of the noise-induced oscillatory multistability in Higgins model of glycolysis, *Chaos*, 28, 033602.

Sandri M. (1996). Numerical calculation of Lyapunov exponents, *Math. J.*, 6, pp. 78-84.

Sargent M., Scully M.O. & Lamb W.E. (1974). *Laser Physics* (Addison-Wesley).

Saucedo-Solorio J.M., Pisarchik A.N., Kir'yanov A.V. & Aboites V. (2003). Generalized multistability in a fiber laser with modulated losses, *J. Opt. Soc. Am. B*, 20, pp. 490-496.

Schawlow A.L. & Townes C.H. (1958). Infrared and optical masers, *Physical Review*, 112, pp. 1940-1949.

Schot H.S. (1978). Jerk: The time rate of change of acceleration, *American Journal of Physics*, 46, pp. 1090-1094.

Schuster H.O. (1988). *Deterministic Chaos* (VCH Weinheim).

Schwartz J.L., Rimault N.G., Hupe J.M., Moore B. & Pressnitzer D. (2012). Multistability in perception: Binding sensory modalities, an overview, *Philos. Trans. R. Soc. B*, 367, pp. 896-905.

Schwartz I.B. & Smith H. (1983). Infinite subharmonic bifurcation in an SEIR epidemic model, *Journal of Mathematical Biology*, 18(3), pp. 233-253.

Seoane J.M., Zambrano S., Euzzor S., Meucci R., Arecchi F.T. & Sanjuán M.A.F. (2008). Avoiding escapes in open dynamical systems using phase control, *Phys. Rev. E*, 78(1), 016205.

Shinbrot T., Grebogi C., Ott E. & Yorke J.A. (1993). Using small perturbations to control chaos, *Nature*, 363, pp. 411-417.

Sprott J.C. (1994). Some simple chaotic flows, *Phys. Rev. E*, 50(2), R647-R650.

Sprott J.C. (2003). *Chaos and Time-Series Analysis* (Oxford University Press).

Sprott J.C. (2010). *Elegant Chaos: Algebraically Simple Chaotic Flows* (World Scientific, Singapore).

Tikhonov A.N. (1948). On the dependence of solutions of differential equations on a small parameter, *Mat. Sb. N.S.*, 31, pp. 575-586.

Van Wiggeren G.D. & Roy R. (1998). Communication with chaotic lasers, *Science*, 279, pp. 1198-1200.

Varone A., Politi A. & Ciofini M. (1995). CO_2 laser dynamics with feedback, *Phys. Rev. A*, 52, pp. 3176-3182.

Volterra V. (1926). Variazioni e fluttuazioni del numero d'individui in specie animali conviventi, *Mem. Acad. Lincei Roma*, 2, pp. 31-113.

Volterra V. (1931). Variations and fluctuations of the number of individuals in animal species living together, In *Animal Ecology*, ed. R.N. Chapman (McGraw-Hill, New York), pp. 409-448.

Wasow W.R. (1965). *Asymptotic Expansions for Ordinary Differential Equations* (Wiley-Interscience, New York, NY, USA).

Weiss C.O. & Brock J. (1986). Evidence for Lorenz-type chaos in a laser, *Physical Review Letters*, 57(22), pp. 2804-2806.

Wolf A., Swift J.B., Swinney H.L. & Vastano J.A. (1985). Determining Lyapunov exponents from a time series, *Physica D*, 16, pp. 285-317.

Xu W. & Cao N. (2020). Jerk forms dynamics of a Chua's family and their new unified circuit implementation, *IET Circuits Devices Syst.*, 15, pp. 755-771.

Yang J., Qu Z. & Hu G. (1996). Duffing equation with two periodic forcings: The phase effect, *Phys. Rev. E*, 53, pp. 4402-4413.

Zagaris A., Gear C.W., Kaper T.J. & Kevrekidis Y.G. (2009). Analysis of the accuracy and convergence of equation-free projection to a slow manifold, *ESAIM Math. Model. Num.*, 43, pp. 757-784.

Zambrano S., Brugioni S., Allaria E., Leyva I., Sanjuán M.A.F., Meucci R. & Arecchi F.T. (2006). Numerical and experimental exploration of phase control of chaos, *Chaos*, 16(1), 013111.

Zambrano S., Marino I.P., Salvadori F., Meucci R., Sanjuán M.A.F. & Arecchi F.T. (2006). Phase control of the intermittency in dynamical systems, *Phys. Rev. E*, 74, 016202.

Index

www.ingramcontent.com/pod-product-compliance
Lightning Source LLC
Chambersburg PA
CBHW050631190326
41458CB00008B/2222